The Rise and Fall of the National Atlas in the Twentieth Century

The Rise and Fall of the National Atlas in the Twentieth Century

Power, State and Territory

John Rennie Short

A
ANTHEM PRESS

Anthem Press
An imprint of Wimbledon Publishing Company
www.anthempress.com

This edition first published in UK and USA 2024
by ANTHEM PRESS
75–76 Blackfriars Road, London SE1 8HA, UK
or PO Box 9779, London SW19 7ZG, UK
and
244 Madison Ave #116, New York, NY 10016, USA

First published in the UK and USA by Anthem Press in 2022

British Library Cataloguing-in-Publication Data
A catalogue record for this book is available from the British Library.

Library of Congress Control Number: 2024932730

ISBN-13: 978-1-83999-247-6 (Pbk)
ISBN-10: 1-83999-247-6 (Pbk)

This title is also available as an e-book.

CONTENTS

LIST OF FIGURES

LIST OF TABLES

ACKNOWLEDGMENTS

I first started the research for this book an embarrassingly long time ago. I started working on the idea of the book 20 years ago. Yes, that long ago. I pursued the matter more intensively, when in 2009 I was awarded a Helen and John S. Best Fellowship to work on national atlases at the American Geographical Society (AGS) Library at the University of Wisconsin Milwaukee. During my time there, the then curator Chris Baruth made me feel most welcome. It was a pleasure to spend a summer month in the AGS Library. It was a good start. But more research was needed. When I returned to my home in Maryland, I started to visit the Library of Congress (LOC) map collection more often than my previous occasional visits. The LOC is not only one of the bibliographic wonders of the world; it is an unparalleled cartographic treasure trove. The Geography and Map Division holds the most comprehensive collection of atlases anywhere in the world, more than 53,000 atlases. It is a rich treasure house for cartographic research. My occasional visits were all too rare as the teaching and other research activities took up my energies. Anyway, I loved the archival research so much that I was in no hurry to finish.

We moved to Washington, DC, in 2013. There were many reasons for the move and the final choice was based on many factors, but not the least was the location. Our new home was only a brisk 15-minute walk from the LOC. I could and did visit more often. Yet the cartographic riches of the LOC meant that there was always one more atlas to consult, one more edition to check. My privileged position allowed me to keep researching. The research, as I came to realize, was becoming less a means to an end and more of a pleasurable experience in its own right.

The staff at the Geography and Map Reading Room was enormously helpful. I am very grateful to the support and encouragement shown by previous chiefs of the division, John R. Hebert and Ralph Ehrenberg. The staff not only met my bibliographic requests but also made suggestions and provided advice. Edward Redmond was always suggesting another edition of a volume when a specific call number proved elusive. Anthony Mullen shared his knowledge of Spanish language atlases and guided me on the path of look-ing at the nineteenth-century national atlases of Mexico, Peru and Venezuela as little-known forerunners of the European national atlases of the twentieth century. Ryan Moore freely shared his encyclopedic knowledge of an early Polish atlas. One afternoon, as I pored over the Eugeniusz Romer 1916 atlas of Poland, he very graciously guided me through the text in front of us for over an hour, sharing his deep knowledge of the text.

At the invitation of Tom Sander, I gave a talk at to the Washington Map Society in February 2016 on my research. A version of the talk was printed later that year in the Society's journal, *The Portolan*. The presentation and published essay were opportunities to gather my research notes and structure a narrative in lecture and print form.

The COVID-19 pandemic halted my visits to the LOC. Stuck at home during the pandemic I was more fortunate than most as it meant that while my archival research was stymied, I could devote more time writing up my research.

In turning the research into this text more debts were incurred. Megan Greiving at Anthem Press was an early supporter of the project and answered my many queries about illustrations with remarkable calm. I am very fortu-nate to have received tremendous support from my institution, the University of Maryland Baltimore County (UMBC). It came in a variety of forms. My research assistant, Abbey Farmer, carefully read through the manuscript more than once and helped with the index. To meet the costs of publication I received generous support from a variety of funds. I received a UMBC Center for Social Science Scholarship Small Research Grant. I also received finan-cial support from two other UMBC sources, the Dresher Center's Scholarly Completion Fund and the College of Arts, Humanities and Social Science Dean's Research Fund. The director of the School of Public Policy, Professor Nancy Miller, was a steadfast supporter of my research and scholarship, over many years. Thank you, Nancy, for everything.

I have also been fortunate with anonymous reviewers. The publisher sent out both the proposal and a completed text to reviewers who made many good suggestions. The two reviewers of the full manuscript deserve a special thanks for a careful and sympathetic reading that improved the text.

I have drawn upon the visual material of a wide variety of atlases to pro-vide the illustrations in this book. It has proved enormously difficult to obtain copyright permissions to reproduce the images that I photographed dur-ing my archival research at the AGS Map Library and LOC. The national atlas is a complex text with multiple authors, and a diverse set of institu-tions are involved in its production and reproduction—some of them only for the duration of the project. Even for the more recent atlases it proved impossible to identify let alone contact the original copyright holders. So, this is an acknowledgment of my reliance on the work of others to produce the images used in this book. I will be delighted in subsequent editions to provide acknowledgment of all copyright holders who come forward.

I think it was Paul Valery, who noted that books are never finished, they are merely abandoned. For many the leave-taking is a source of relief, even joy. Not in my case. The book you have before you was reluctantly finished. Rather than relief, I have a sense of loss. I can only hope that my loss is your gain.

Chapter 1

INTRODUCTION

A useful working definition of a national atlas is "a generally comprehensive, officially sanctioned single-country atlas."[1] The publication of *Atlas öfver Finland* (*Atlas of Finland*) in 1899 marks the beginning of the modern national atlas since it has all the main attributes of subsequent national atlases produced over the course of the next 100 years (see Figures 1.1 and 1.2). These include

- a comprehensive and official, or officially sanctioned text.
- a symbol and embodiment of national identity.
- a tool for government to inventory, classify and depict the national territory.
- a text aimed at multiple and extended audiences including the international scientific community as well as a domestic readership.
- a part of the ideological apparatus for education into, and promotion of, citizenship.
- a display of the biopolitics of the state in its depiction and classification of the population.
- a depiction of national space that also makes global connections.

The heyday of the national atlas coincides approximately with the twentieth century.[2] The modern national atlas mirrors and embodies some of the important themes of this turbulent century, including the complex connections between nation, state and territory; the rise of state-sponsored science; the full emergence of biopolitics; the active creation of a national identity; and the development of mass literacy and state education, in general and cartographic literacy in particular. Nation-states did not simply emerge. They were actively created and managed and the national atlas was an integral part of nation-making. The rise of the modern national atlas and its changing form provides an intriguing window into the connections between nation-state, science, territory and power.

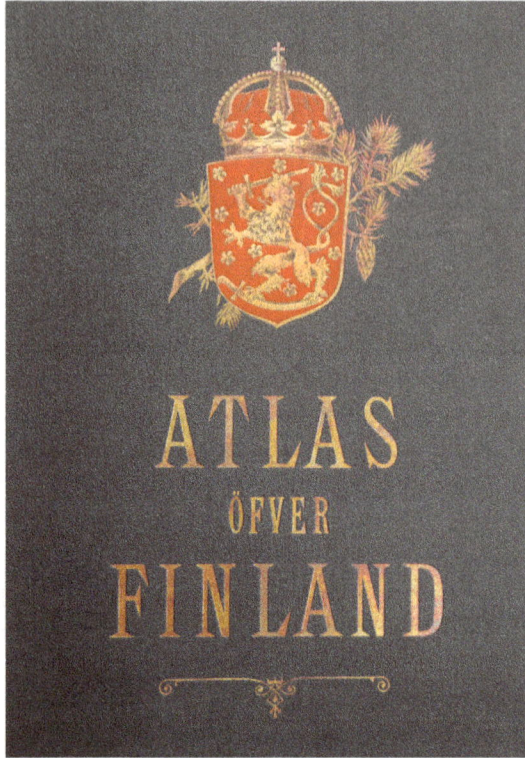

Figure 1.1 Cover, *Atlas öfver Finland* (*Atlas of Finland*), 1899. *Source:* Photo by John Rennie Short.

The national atlas is a complex text. It is a scientific document, a tool of legitimation, a spatial claim, a textual appropriation of territory and an attempt to foster, encourage and create a national community. The national atlas is an attempt to represent and legitimize the territory of the state for both an internal and external audience; it is an element in national ideological constructions encased in global scientific discourses. I aim to unfold these complex layers of the national atlas.

Between 1900 and 2000, more than seventy countries produced a national atlas, an official or quasi-official rendering of the nation-state in maps and accompanying text. I consider the reasons behind and the characteristics of this state-sponsored cartographic explosion. The primary material for this study is a close reading of the national atlases of 40 countries from across the world. They are shown in Table 1.1 and a more detailed bibliography is provided in the Appendix. In many cases I looked at multiple editions of the same national atlas to ascertain changing assumptions and shifting

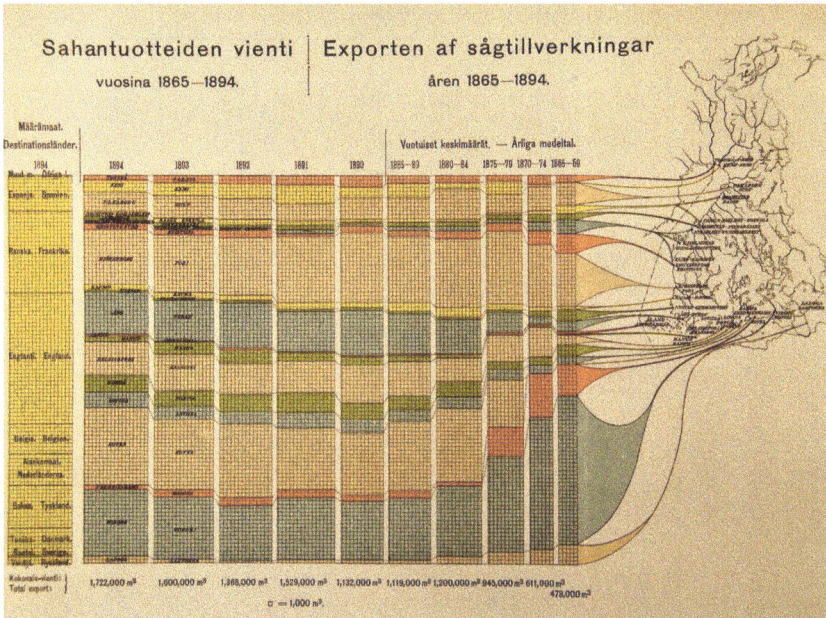

Figure 1.2 Diagram, *Atlas öfver Finland* (*Atlas of Finland*), 1899. The text, in both Finnish and Swedish, refers to the export of timber products from sawmills to different countries. *Source:* Photo by John Rennie Short.

discourses. This is not a sample in the strict statistical sense but in the more general sense of a wide range of countries from rich to poor, progressive to regressive and capitalist to (at the time) communist.

A Century of Nationalism

The twentieth century was a century of nationalism. As old empires collapsed, new nation-states emerged in revolutionary ruptures and postcolonial emergences to create a patchwork of fledgling nation-states across the globe. In 1900, there were only 57 independent countries, 70 by 1930, and by 2000, the United Nations had 189 members.

The world splintered as old empires collapsed and were dismembered. The Austro-Hungarian Empire, for example, splintered after the end of World War I into Austria, Bosnia, Croatia, Czech Republic, Hungary, Italy, Montenegro, Romania, Serbia, Slovakia and Ukraine. Rapid decolonization after World War II meant that former colonies such as Burma, Cambodia, Kenya, India, Pakistan and Vietnam achieved independence and became nation-states. The collapse of Yugoslavia after 1989 resulted in six new

Table 1.1 National Atlases.

Afghanistan	1985
Argentina	1886, 1954, 1962
Botswana	2001
Brazil	1966, 2000
Canada	1905, 1906, 1916, 1958, 1974, 1987–93
China	1987, 1994, 1999, 2000
Colombia	1906, 1967, 1992, 2008
Cuba	1884, 1898, 1949, 1970, 1978, 1979, 1989, 1991
Estonia	1925–28
Ethiopia	1988
Finland	1899, 1910, 1925, 1960, 1976–95
Ghana	1928, 1945, 1970
India	1959, 1968, 1979, 1982, 2003–9
Iran	1973
Israel	1956, 1970, 1985, 2011
Jamaica	1971, 1989
Japan	1943, 1977, 1990
Kenya	1959, 1962, 1970, 2003
Korea	2009
Latvia	1938
Liberia	1983
Lithuania	1918–29
Mexico	1858, 1885, 2002
Pakistan	1985, 1997, 2012
Peru	1865
Poland	1916, 1947, 1973–78, 1993
Qatar	2006
Romania	1974
Russia	2004
Saudi Arabia	1975, 1999
Senegal	1925, 1977
Sierra Leone	1953
South Africa	1960
Spain	1965, 1992
Sri Lanka	1988, 2005, 2012, 2016
UAE	1993
USA	1874, 1883, 1898, 1903, 1914, 1925, 1970, 1988
USSR	1937–39
Venezuela	1840, 1979
Vietnam	1909, 1928, 1996

independent countries. The fall of the Soviet Union in 1991 in turn led to the creation of 15 new states.

The Main Arguments

These new states as well as the old ones had to be represented, explained and legitimized. One vehicle was the national atlas, a published collection of maps that represented the territory and the people. I draw upon and extend five theoretical arguments: *cartographic state, social deconstruction of the map, imagined communities, national imaginaries* and the nexus of *nation-state-territory*. These five arguments are not so much separate lines of inquiry as they are different facets of the subtle connections between the cartographic power of the state and the use of the national atlas to generate a national territorial imaginary, create and reinforce national identity and occupy a space in an international discourse of science.

The first is the notion of the *cartographic state*, a term developed by Jordan Branch.[3] He suggests that since the early modern period, maps play an important role in how rulers imagine and rule their territories. States are in effect territorial claims embodied in a mapped linear boundary. The cartographic representation of this territory is full of political significance and social meaning. Many of the early modern states were just as much cartographic representations on paper and vellum as they were political practices on the grounds or as sources of collective identity. Modern statehood emerged as much on maps as in the minds of the populace. States were, and continue to be, defined by their territorial depiction. By mapping a territory, the state reinforces its claim to power, dominance and legitimation. The surveillance of territory was and remains—through maps and increasingly in CCTV, drones and satellites—a vital element of state power.

This book considers one of the most explicit spatial practices of the modern cartographic state, the *national atlas*. Making an atlas is not only a scientific enterprise but also a rhetorical act, a discursive strategy and a territorial claim. A national atlas is a cartographic depiction of territory and space that the state claims to own, control, manage, explain and describe. It is deeply bound up with the obsession with representation as a form of accurate comprehension and knowledge consumption. The national atlas is akin to what Timothy Mitchell refers to as the "world as exhibition" form of control—in this case, providing new conceptions of national and international space.[4]

The second argument is what we may term the *social deconstruction of the map*. In a landmark book, published in 2001, Brian Harley shifted the emphasis in the history of cartography away from the idea of maps as simply technical documents that recorded scientific progress.[5] The history of cartography in the traditional interpretation was as an inexorable march toward "enlightenment," as unknown land became known and the "gaps" in the world map were filled in. Harley, drawing on the work of Foucault, argues that maps and mapping are documents marked by social and political power. Maps, and by extension atlases, are not innocent scientific texts but arguments with an agenda that marginalize some, exclude others and seek to render "natural" that which is "social." The map is not a mirror of nature but a social construction. In this postmodern perspective, maps can be used to reveal the presence and absence of power. In the following chapters, I explore the national atlas as a social construction that reveals the social power at work in the representation, understanding and management of national territory.

Third, in an argument of great subtlety Benedict Anderson argued that nations are not so much simple facts of race or ethnicity but rather what he terms *imagined communities*.[6] He paid particular attention to the role of print capitalism in creating a national discourse. Anderson identified three institutions of power in the *census*, the *map* and the *museum* that together, respectively, allow the state to imagine the people under their dominance, the geographic territory under their control and the nature of their historical legitimacy. In this book, I want to extend his idea of the *map* to include the national atlas where collections of maps in one volume play a vital part in the justification of state legitimacy, the creation of a national community, the presentation of the nation to the global community and the insertion of the state into global discourses.

A national identity is fostered, encouraged and created by a shared cartographic understanding of the nation. A state is embodied and understood in its cartographic depictions and a national community can even be defined by the widespread usage of its cartographic convention. The saturation of cartographic images has created a widely accepted semiotics of states. Outline maps of the United States or of Italy or of the United Kingdom, for example, are easily recognizable: they are used as symbols of these nation-states because of their quick and easy recognition/understanding.[7] Maps of a nation-state are not just depictions of surface area or even of representations; they also embody the nation-state. The geo-body of the nation, the concrete identification of national territory, is discursively created through the mapping of territory.[8] The national atlas is one of the most concrete depictions of a national territory.

A national atlas is more than just a collection of maps of national territory. It is also an important element in a suite of active nationalisms that embody and celebrate national identity. "National" events are enacted and reenacted, "national" stories are told and retold and "national" identities are created through commemorative activities, exhibitions, fairs and sacred sites. The national atlas is an active embodiment of the territory, peoples and culture of a nation-state. Whereas Anderson made a distinction between census, map and museum as instruments of state power, I will show that national atlas combines elements of all three. They not only show the territory under control but also the people under its dominance and the nature of its historical legitimacy.

Fourth, we can further Anderson's notion of *imagined community* by exploring the idea of *national imaginaries*. These are one element in the taken-for-granted geographical imaginaries that connect space and society, identity and place, and meaning and territory into active, constantly changing constructs that move back and forth from representations of reality to effective constructors of reality. Friedrich Nietzsche noted that imaginaries encompass more than the real world: there are no facts, only interpretations and, in his widely cited but impossible to source quote, that "whichever interpretation prevails at a given time is a function of power and not truth." Imaginaries fill out the world with hopes and wishes as well as with facts and observation: they describe and explain as well as affect and influence. Imaginaries embody, shape, inform and condense power relations into geographical worldviews.

National imaginaries include *national environmental ideologies* that reference how myths of wilderness, countryside and city are expressed in the national territory.[9] The national atlas is a particular form of national imaginary that is informed by the discourses of science and nationalism, rational planning and emotive commitment, and globalism and national distinctiveness. These discourses are, at one and the same time, scientific artifacts as well as ideological expressions of hopes and fears.

So far, I have been using the terms "nation" and "state" almost interchangeably. But the reality is more complex. The fifth idea that I consider in this book is the complex space of nation-state-territory. Let us look at some definitions. The term "nation" implies an aggregation of people sharing similar race, ethnicity, history or language. Earlier usage stressed race while more recent usage may imply race but makes broader appeals to shared identity and history. A nation implies shared characteristics and by extension shared interests. The nation is a geographical imaginary that connects a people with the occupancy of place. The "state," on the other hand, is the government by political authority: it has or seeks monopoly power over an area of territory. Both nation and state have a territorial component. Nations are generally considered contiguous aggregates while the state is a spatial entity defined by the area under its formal authority. But there is no simple territorial connection between nation, state and territory. There are nations without states and states with more than one nation. The national identity of many states is rarely simple or uncontested, and different nationalities often fit awkwardly into the spatial boundaries of the state. The spatial unit of the state may contain more than one nationality. "Territory" generally is considered to designate an area of the earth. Yet its original Latin—*terrere*—means "to frighten" and has been transformed over the centuries to designate a place from which people are warned off.[10] The multiple questions that arise from this linguistic thicket of nation-state-territory include the following: Who are the owners? Who is in control? Who is to be warned off and how? The national atlas gave spatial expression to the complex territoriality of nations and states. In this book, I will discuss how the territorial ambiguities of nation-state-territory are explored and ignored, excavated and elided in the national atlas.

Structure of the Book

Chapter 2 looks at the forerunners of the modern national atlas. There are only a few examples of what we may term the premodern national atlas. Timothy Pont produced detailed manuscript maps of Scotland between 1583 and 1614; it is an incomplete coverage of the nation with enigmatic origins. The national atlas in its embryonic forms appears in sixteenth-century England and France with Christopher Saxton's 1579 *Atlas of England* and Maurice Bouguereau's 1594 *Theatre françois*. Both were produced during difficult times for the respective monarchies. Both countries were riven by murderous religious factionalism and the corresponding fear of foreign interventions. These two early atlases depict the territory of the nation as part of a state surveillance during rebellious times. But the national atlas for general consumption, the modern national atlas, only emerged in the nineteenth century with the rise of nationalism, print capitalism and greater literacy. In Chapter 3, "Cartographic Anxieties and the Emergence of the Modern National Atlas," I argue that an early form of the modern national atlas emerged from a "cartographic anxiety." This was most pronounced in Latin America where new states appeared from continental-wide imperial territories. Printed at a time of limited literacy and of expensive publishing, their purchase and readership were restricted to an elite audience.

National atlases did not develop from a vacuum as free-floating entities. They emerge tethered and connected to existing cartographic practices and traditions. In some cases, the break was sharp and sudden and in other cases more gradual, but all cases were connected to what went before. The modern atlas of the twentieth century arises particularly in the context of ruptures, including the postcolonial, the newly independent and the recently invented. In Chapter 4, "Cartographic Ruptures and the National Atlas," I look at how national atlas emerged from revolutionary ruptures and postcolonial reorganizations. I will demonstrate how new countries emerged from these imperial breakups and national reorganizations—post–World War I and II—and how the decolonization that began in the 1950s led to the reimagination of the national atlas. In some cases, the prerupture and postrupture atlases will be compared. I will show how states used the national atlas to build national identity and construct national coherence.

In subsequent chapters, I will excavate various discourses and silences within the national atlases. Chapter 5, "National Atlas, Global Discourses," looks at how the national atlas portrays the nation in a wider international setting through the languages used and the map projections employed. I will show how the national atlas was an important element in the development of a national scientific community linked to a growing global discourse of science. The atlas was a product of scientific communities, both national and international.

Chapter 6, "The Physical World of the National Atlas," focuses on the discourse of physical sciences. I look at the coproduction of science and the state, the origins of the science of the national atlas and the contested discourses of the national atlas, paying particular attention to boundaries, toponymic issues and changing understandings of the physical nature of the territory. An examination of different editions of a national atlas reveals changing environmental discourses of how nature is commodified, represented and studied. I will show how the physical environment was re-represented from an inert container to be exploited to a place of environmental hazards and sustainable growth to a living organism that had to be carefully managed and protected. In Chapter 7, "The Social World of the National Atlas," I examine how human populations were presented and represented, how the state dealt with alternative readings of the nation-state and to what extent did atlases incorporate subaltern or alternative cartographies that contest, question or disperse forms of power. Chapters 6 and 7 provide an opportunity to look at how the different discourses reveal new and persistent themes. National imaginaries are rarely coherent, consistent or stable. I will look at successive editions of the same national atlas to plot some of these changes. In the final chapter, I look at the decline of the modern national atlas, especially in its traditional textual form, and hint at new emerging forms.

The overall aim of the book is to use the national atlas as a novel way to reveal the connections between power, state and territory in the twentieth century. The national atlas showed not only the territory under the state's control but also gave it a scientific rationale and historical legitimacy. It was a cartographic representation that fused the state and the nation, the people and place, government and citizenry, national history and geography, and represented the distinctly national global discourses. I will show how states used the national atlas to embody the country in textual-cartographic form, giving it a territorial container, a discursive identity, a spatial shape, a history, a geography and a statistical reality.

Chapter 2

THE EARLY NATIONAL ATLAS

The national atlas emerged with the rise of the nation-state in the sixteenth century. It was primarily a tool of surveillance connected to the need for a more spatially informed statecraft. Sometimes printed, more often in manuscript form, its circulation was limited. It was used to inform the military, economic and political elites. I will explore the case of France and particularly of England.

A Cartographic Explosion

There was a cartographic revolution in Europe in the sixteenth century. In 1500, maps were almost unknown and little understood; by 1600, they were familiar objects and vital parts of national life as a widespread cartographic literacy emerged. The culmination of this cartographic revolution was an incredible burst of cartographic innovation in the last third of the sixteenth century. In a relatively short period, the first modern atlas (1570), the first urban atlas (1572) and the first national atlas (1579) were all produced in Europe. Within the space of less than 10 years, the standard representations of the world that we still make use of today were first developed.[1]

In 1570, Abraham Ortelius (1527–1598) published one of the world's first atlases, the *Theatrum Orbis Terrarum*. At that time, most maps—especially the larger maps—were single sheets kept rolled up and unrolled every time they were used. There had been a collection of maps brought together in one volume before Ortelius, but he was to set a standard by which subsequent collections would be judged and compared. Ortelius was not only concerned with providing maps in the handier book form, but he also wanted the very best maps. He sought out the most accurate current maps and collected maps from a wide range of cartographers. He listed all the names of the cartographers whose work he had used: 87 names were mentioned for the first edition in 1570, 170 by 1595 and 182 in the 1603 edition.

These world atlases were part of early modern Europe's attempt to describe and picture a new world opening up to market penetration and imperil designs. European global territorial appropriation was matched by a visually rich epistemic project to represent, classify and cohere this new world. The Dutch, and especially Dutch atlas makers, played an important role in this endeavor to not only describe but also to appropriate this world.[2]

By the last third of the sixteenth century, there was a considerable stock of urban maps and images. This stock had been drawn for a variety of reasons:

- civic pride,
- celebrations of specific events, such as the colossal prospect of Cologne by Anton Woensam drawn in 1531 on the election of Ferdinand of Austria as king of the Romans,
- military surveillance,
- parts of national inventories.

Compilations of city maps and prospects were published in 1551 and 1567, but the first city atlas was the *Civitates Orbis Terrarum* by Georg Braun and Frans Hogenberg. One volume was published in 1572, but it became so popular that by 1617, the work consisted of six volumes with over 363 urban views.

The national atlas was an important part of this cartographic revolution. Timothy Pont, for example, produced detailed manuscript maps of Scotland between 1583 and 1614.[3] But for printed maps we must look to England and France and especially to Christopher Saxton's 1579 *Atlas of England* and Maurice Bouguereau's 1594 *Théatre françois*. Both were produced during difficult times for the respective monarchies. Both countries were riven by murderous religious factionalism and the corresponding fear of foreign interventions. These early printed atlases depicted the territory of the nation as part of a state surveillance during rebellious times. Printed at a time of limited literacy and expensive publishing, their readership was restricted to an elite audience.

The Early National Atlas in England and France

Christopher Saxton created the very first national atlas in Europe in 1579. He was born around 1542/44 in Yorkshire. We know little about the details of his early life. He became a surveyor. The dissolution of the monasteries had created a large pool of commodified land as the vast estates of the religious orders became part of the commercial land market. The enclosure of land—a process of privatization of public lands into private hands, especially into the hands of the already wealthy—was creating a more capitalistic land market that needed to be mapped and surveyed. Saxton was employed as an estate surveyor by private landowners and by official courts investigating land ownership disputes. Saxton became well connected and prospered in the shifting quagmire that was Elizabethan England. In 1573, he was chosen by Thomas Seckford to survey and map the counties of England and Wales. Seckford was a state functionary, master of the Court of Requests and later Surveyor of the Court of Wards. His boss was William Cecil, later Lord Burghley, who was one of the most powerful people in the land, the Queen's Secretary of State, a Privy Councilor and later Lord Treasurer. He was concerned with surveying the realm for many reasons, prime being the need for accurate and regular surveillance at a time when religious factionalism split the country and English Catholics looked to Spain for support to overthrow Elizabeth's Protestant regime.

Saxton was officially appointed by the Queen to undertake the survey of the country in July 1573. Seckford paid all of Saxton's costs, but it really was a state project promoted by Burghley. In return, Saxton received official favor. In 1573, he was given an estate in Suffolk, formerly monastery lands. In 1574, he got another grant of land in London and in 1579 he was given a formal coat of arms. His crest was a hand and arm holding a pair of partly open compasses.

Saxton's survey lasted from 1574 to 1578. He began in Norfolk and the south Midlands, then moved to Essex, east Midlands and then into the north. All the English counties were completed by 1577 and all the Welsh counties by 1578 (see Figure 2.1).

Saxton was given an open letter to local officials directing them to take him to a high place to survey the land. The aim was not so much to identify latitude and longitude, which do not appear on his atlas maps; it was to identify the lie of the land. Saxton's maps surveilled the national territory at a time when Catholics were feared to be plotting against the Queen. In 1570, the pope, in a declaration that can only be described as a form of Catholic *fatwa*, had given a free pass to heaven to any Catholic able to kill Elizabeth. The Spanish posed a continual threat. In 1567, the Spanish army in the Netherlands was dangerously close. In 1569, the Privy Council called on counties to have a general muster of all able-bodied males aged over sixteen and to create a beacon system of bonfires on hills to alert of foreign invasion. The beacon sites were probably the high points that Saxton was taken to by the local officials. The route of Saxton's survey reflected strategic considerations. He surveyed the vulnerable south coast counties before the less strategic northern counties and Wales. The very use of counties as the principal unit reflected the chain of command. Counties were a military unit of allegiance to the Crown: the power flowed from the Crown to the Lord Lieutenants of the county to local justices of the peace. The counties were responsible for military musters. Counties were not the rather quaint

Figure 2.1 Cheshire, *Atlas of England and Wales*, 1579. *Source:* https://upload.wikimedia.org/wikipedia/en/d/d8/Cheshire-south-east-saxton-map-1577.jpg

divisions they are today; they were the principal units in the spatial, military and political administration of the realm. Saxton used this basic military-political unit as his template.

Having surveyed all the counties of England and Wales, Saxton brought them together in an atlas. Burghley took an active interest in Saxton's work. He saw each plate as soon as it was engraved and made them up into his own atlas to which he added notes and other maps. This atlas received lots of annotations leading up to and during the attempted invasion of England by the Spanish Armada in 1588. On Saxton's map of Northumberland, now in the British Library, Burghley annotated in his spikey handwriting the number of horses that the local gentry could raise in the event of war.

The 1579 *Atlas* has a frontispiece of Elizabeth, showing her as the patron of geography and astronomy. The first map is a map of England, followed by double-paged maps of the individual counties. The maps show rivers, towns, enclosed forests and the houses and castles of local gentry. Relief is shown in diagrammatic form as little molehills, roughly to scale. No roads are shown, suggesting that the survey was done quickly and rather crudely. Roads were probably invisible from the highpoints and their accurate mapping would have taken up too much valuable time. The county maps in the 1579 *Atlas* were a vital geographical intelligence. There was no text to accompany the maps. However, the stamping of the royal coat of arms on each and every map speaks volumes about the connection between the body of the country and the figure of the royal personage. In the famous portrait of Elizabeth (c. 1592) by Marcus Gheeraerts the Younger, the Queen is depicted standing on a map of England and Wales, a map of the country/body of the monarch in one compelling image (see Figure 2.2).

Very much a product of its time, the atlas was to become a landmark in English cartography. It was published throughout the seventeenth century and was the basis for most county maps until 1650. Saxton provided the basis for English maps for almost one hundred years and provided the most complete survey of the country until the creation of the Ordnance Survey in 1791.

Figure 2.2 Queen Elizabeth I ("The Ditchley portrait") by Marcus Gheeraerts the Younger oil on canvas, c. 1592. *Source:* © National Portrait Gallery, London.

Saxton's work atlas is a good example of an early national atlas: an official project primarily designed as a form of territorial surveillance and designed for an elite audience. It was not meant to legitimize the state but to provide information for the political elite.

The first national atlas of France, *Théâtre françois*, was published in Tours in 1594. It was a compilation of maps from a variety of sources collated together and dedicated to King Henry. The various uses of the atlas were cited as educational military intelligence, tax purposes and trade. In the first edition, the picture of Henry covered the map of France; flick back the portrait and there, underneath, is the map of France: king and country, nation and sovereign, the body of the country and the head of the king. The iconography is clear and direct. The country was detailed in 48 maps, many of them with inserts of town plans.

These early national atlases also played important roles in the creation of a fledgling national identity. Nationalism is a social creation not a biological given. Many things go into the production of nationalism, including schooling, religion, rituals, stories as well as holidays, history and geography. One essential element was, and continues to be, a recognizable image of the country. The two processes of cartographical representation and emerging national identity are closely connected. The maps of the sixteenth century that are integrated in national atlases are not just depictions of places; they are also generators of national consciousness. This was never a simple process. The national atlases could be and were read in very different ways—signs of absolutist power as well as indicators of a wider, broader national community. To map the country was not a technical exercise, devoid of political meaning. When the country was mapped and, as in the case of England, individual counties brought together in one volume, national identity was given cartographic form, a physical and textual presence. The early national atlas was primarily an administrative/surveillance device that also embodied the nation-state and reinforced national consciousness. A project designed to survey the nation also helped to make the nation.

Let us end this chapter of the early national atlases by noting their echo in the more contemporary national atlases. The way that the image of King Henry in the *Théâtre françois* used to reinforce the connection between the ruled and the ruler is also employed in the embossed heads of Lenin and Stalin in the title page of the 1937 volume of *Bol'shoĭ sovetskiĭ atlas mira* (*Great Soviet World Atlas*); in the formal photographs of the shah, his Queen and Crown Prince in the 1973 *Atlas of Iran White Revolution*; of Vladimir Putin in the 2004 *Nat͡s ional'nyĭ atlas Rossii v chetyrekh tomakh* (*National Atlas of Russia*) and of the emir and his heir apparent in the 2006 *al-Aṭlas al-waṭanī al-Qaṭarī* (*Qatar National Atlas*). In some countries, the modern national atlas still embodies the codepiction of territory with images of its rulers.

Chapter 3

CARTOGRAPHIC ANXIETIES AND THE EMERGENCE OF THE MODERN NATIONAL ATLAS

What we would now recognize as the modern national atlas first begins to appear in the nineteenth century. In the early and mid-nineteenth century, the national atlas emerges as part of the creation of new nation-states. It was used to create a cartographic panopticon that informed the elites, to attract global investment, to generate a national cartographic literacy and to legitimize the new nation-state. Later, the national atlas came to fully flower in the twentieth century with mass education and literacy, cheaper printing and a greater need to legitimize popular support. The modern national atlas began and remains a tool of surveillance but, as the twentieth century progressed, developed into an important element in the state's ideological apparatus.

States rarely emerge fully formed. Instead, states often arise from a liminal space of boundary uncertainty, which gives rise to a *cartographic anxiety* which we can define as a fear for the integrity of the body politic of the state and the stability of the geo-body of the nation. It is expressed by national maps that seek to reaffirm the security and indeed the inviolate nature of national boundaries. The term was first employed by Sankaran Krishna who used it in reference to postcolonial society in general and to India in particular.[1] Cartographic anxiety is most acute in the definition of boundaries—both the physical and the social—in postcolonial societies, emerging from territorial ambiguity and boundary uncertainty. I will discuss three examples of national atlases marked by the cartographic anxiety of a postcolonial Latin America.

The Modern National Atlas in Latin America

National atlases often emerge as part of the creation of new nation-states. Several atlases prefigure the 1899 *Atlas öfver Finland*. In South America, local Spanish-speaking elites resisted and finally overthrew Spanish domination.

As part of their new national identity, intellectual elites worked to create a national atlas that not only would reflect and embody rising nationalist sentiment but that also would function as a scientific document to link the new nation to the global Enlightenment project. The national atlas became an integral part of the Enlightenment project that shared the Humboldtian vision of understanding and representing the world in a discourse of rational thought and precise measurement. Produced at a time before mass literacy, these proto-modern atlases were limited to an elite readership and so played little role in the creation of widespread national identity.

This process began in South America where in the wake of independence, after the overthrow of Spanish colonial control, new national identities and new national atlases were required. In addition, the national atlases were produced to further accommodate territorial claims in a time of uncertain and contested sovereignty; as scientific documents to link the new nations to the global Enlightenment project; and as repositories of information to guide and encourage investment, immigration and trade.

The context

From the sixteenth century, Spain (and Portugal) gained vast territorial possession in Central and South America. All real power was centralized in Spain. Colonial local economies were structured for Spain's enrichment. Trade with foreign countries was forbidden and only the Spanish-born could plant grapes, grow tobacco or own a mine. Manufacturing was banned and punitive taxes made their way back to Spain with little local investment. Those with Spanish ancestry but born in the New World, especially chafed at these political and economic restrictions. They were denied access to the highest levels of political office reserved for those born in Spain. Even their economic power as large landowners was limited by taxation by Spain and trade restrictions that tied them to the Spain imperial system rather than expanding markets of the UK and the United States. The American Revolution (1776) and French Revolution (1780) had inaugurated a new era: distant, absolute power was being questioned and old colonial ties being severed. The invasion of Iberia by Napoleon and the overthrow of the king of Spain in 1808 gave new hope to the local-born elites of Latin America eager to loosen the bonds. Napoleon's resounding victory demonstrated that the imperial center was vulnerable, a spent rather than an ascendant force. While Brazil peacefully separated from Portugal in 1822, it was not so easy for the Spanish possessions. Spain was a failing power eager to maintain its colonies for the wealth and the prestige, especially because its relative position in Europe was deteriorating in comparison with the dynamic colonial and commercial empire of the UK.

The result was a bitter military struggle that began in isolated uprisings immediately after Napoleon's invasion of Spain but grew in the subsequent decade. The insurrections eventually wrested control sway from Spain. The Argentine Republic was declared in 1816 and forms of independence were achieved in Chile (1818), Colombia (1819), Venezuela (1821), Mexico (1821), Ecuador (1822), Peru (1824) and Bolivia (1825). Political independence did not inaugurate political stability. There were no obvious territorial units for the independent countries that emerged. Spain ruled over a vast empire that was breaking up into arbitrary units that did not draw upon a precolonial coherence but on the exigency of revolutionary movements. The result was a period of territorial ambiguity when national boundaries were neither fixed nor agreed upon. There was a territorial reorganizing as new states fought over territory and boundaries.

There was territorial uncertainty caused by conflicts between different groups, competing claims and confederations that would emerge only to break up. In 1821, Gran Colombia included Colombia—encompassing, present-day Panama—as well as Ecuador and Venezuela. But by 1830, the confederation had split into three separate parts. Later, in 1903, Panama, with U.S. encouragement and assistance, broke away from Colombia. In 1823, Central America seceded from Mexico, and only seven years later the breakaway territory further splintered into five separate republics. Fragmentation and territorial restructuring soon followed emancipation from the Spanish Empire. Nation-states had to be invented and created. Postrevolutionary administrations had to untangle land claims, attract investments and create new administrative spatial units. They needed to fix boundaries, facilitate economic development, create administrative spaces and foster social legitimation. National atlases emerge from this confusion of new states with uncertain boundaries and competing territorial claims, the economic need to attract foreign investment and to forge economic and intellectual links with a wider world. New administrations produced the territorialized state through cartographic practices of delineating territory, of fixing boundaries and of assembling administrative units.

The national atlases also emerged at a time of the rise of Enlightenment science with its emphasis on the orderly accumulation of knowledge. It is no accident that the revolutionary leader Simon Bolivar (1783–1830) also communicated with the Enlightenment scholar of the age, Alexander von Humboldt (1769–1859). Bolivar involved in the creation of the six new nations of Bolivia, Colombia, Ecuador, Panama, Peru and Venezuela and the Enlightenment scientist Humboldt, whose work stimulated measurement and mapping as vital forms of scientific fieldwork, met in Paris in 1804 and again in Rome in 1805.[2] The revolutionaries of Latin America laid claim to the wider Enlightenment project to usher in the modern world.

A national atlas had the possibility to be a foundational document reframing territory in a national postimperial space and providing an inventory for national economic development and encouraging foreign investment in an increasingly globalizing economy. But there were many difficulties, including the cost and lack of skilled personnel. An atlas required skilled mapmakers, surveyors and cartographers. So, despite the need there are few examples. I will discuss three.[3]

Venezuela

The 1840 *Atlas Físico y Político de La República de Venezuela* was commissioned in 1830, the same year that the Republic of Venezuela emerged from the collapse of Gran Colombia. One of the new Republic's founding documents is the national atlas.

The new government hired Giovanni Battista Agostino Codazzi (1793–1859).[4] He was born in Lugo, Italy, filled with Republican ideals and served in Napoleon's army. After Napoleon's defeat, he left Europe, ending up in Venezuela where he offered his services to Simon Bolivar. He became a prolific mapmaker at a time when the new republics desperately needed to map and survey the national territory. He was made governor of a region in the southwest of Venezuela. After a military insurrection, he moved to Colombia where he continued his mapmaking and later completed a survey of Panama for the British government. The Panama Canal follows the line he laid out in his survey of 1854. He also put in charge of the nation's military school and mapping authority an institution still called the *Instituto Geografico Agustin Codazzi*. He died of malaria while mapping in the mountains for the ambitious *Comision Corografica*, established in 1850 to create a national inventory of Colombia's infrastructure, resource base, lands, river ways and sites of industrial production.

The new Republic was given cartographic legitimation in the atlas. It contains 30 maps. The physical geography of the country is represented in diagrams of the mountains and rivers and in maps of climate and agricultural zones, hydrography and relief.

The maps are employed to clearly distinguish the new state from the surrounding countries. In the atlas, the state's territory extends east to the Essequibo River and into an area with a notation (see Figure 3.1) that in English translates as "territory usurped by the English." The state is given a solid definitive cartography at a time of fluid and changing territorial boundaries and state formations. Most of the maps are detailed maps of provinces and cantons. The provincial boundaries of the new state are prominent as are maps of the different provinces, in effect creating the administrative geobody of the new state.

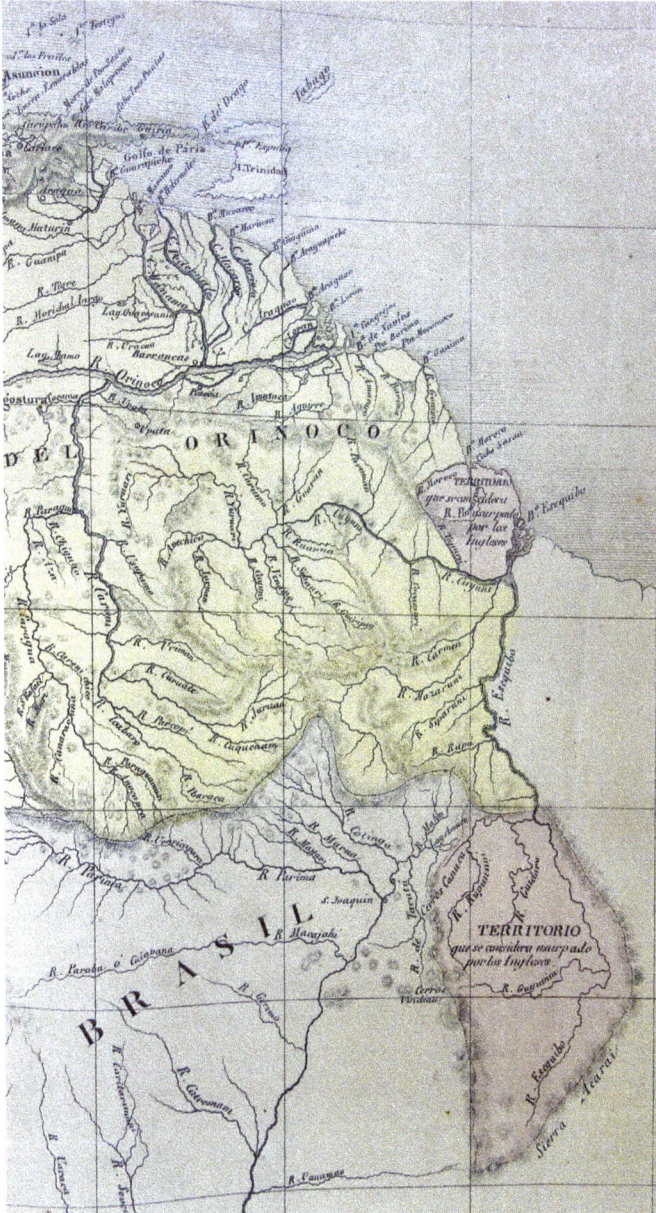

Figure 3.1 Territorial map, *Atlas Físico y Político de La República de Venezuela,* 1840.
Source: Photo by John Rennie Short.

This atlas was arguably one of the first to become part of the flow of geographic-scientific knowledge that was linking national scientific institutes, scientific societies and individual scientists into a global Enlightenment discourse. The atlas was published in Paris and Condazzi sent the atlas to various scientific societies in Berlin, London and Paris as he worked to have the atlas recognized in the emerging global discourse of science. No less than Alexander Humboldt and the Academie de Sciences de Paris praised the accuracy of his atlas.

Peru

Peru achieved independence in 1824. As with other South American countries, there was territorial ambiguity and confusion. A Peru-Bolivian confederation was a short-lived affair after rival neighbors Chile and Argentina invaded during the 1836–39 War of the Confederation. Later, in the 1879–84 War of the Pacific, Peru lost mineral-rich territory to Chile.

The *Atlas Geográfico del Perú* was published in 1865: it is the first atlas of the country. This was a time of relative stability after the War of the Confederation and before the War of the Pacific. Government revenues were huge from guano exports for an expanding world market. However, the atlas was published in France not in Peru with an edition in Spanish and another in French. At the time, most scientific publications of Peru were published in France. The metropolitan pull, even for overtly nationalist projects, was still strong. The editor/author of the atlas was Mariano Felipe Paz Soldan (1821–1886), then director general of public works, who on the title page proudly lists his membership in the Royal Geographical Society of London and the Humboldt Society of Mexico. He was a politician, in office from 1869 to 1870, and an important state functionary and public intellectual. He wrote a history and compiled a geographical dictionary of the country. He suggested reforming the currency, and in 1869 he proposed a telegraph line to North America with the capital to be raised in London.

The atlas contains 45 maps and plans. It is a resource inventory and a territorial organization of the state. There are maps that show the location of mineral reserves, an important part of the Peruvian economy's links with the wider world, then as now. The mineral maps indicate deposits of arsenic, borax, gold, gypsum, iron, lead, saltpeter, mercury and silver. There are maps depicting the boundaries of the state as well as different departments and provinces. There are also pictorial depictions of the country in 23 views and scenes of churches, landscapes and "natives." Developments in printing now allowed more images and pictures to be shown in an atlas, giving a distinctly multimedia dimension to the cartographic text. There are maps depicting the boundaries of the state as well as different departments and provinces. The political element is clear: the nation is shown as a homogenous whole made up of organic, constituent departments.

Soldan wanted to publish a national atlas that would garner international respectability. He saw it as a scientific enterprise, a self-conscious attempt to link with metropolitan centers of science and knowledge production and part of a wider international scientific discourse. One table in the atlas records the latitude and longitude of places as measured by explorer-scientists such as Humboldt. The atlas contains a common late nineteenth-century diagram of depicting the country's mountain structure in a single image. It embodies Humboldt's understanding of biogeography and especially in the relationship

Figure 3.2 Topographical diagram, *Atlas Geografico Del Peru*, 1865. *Source:* Photo by John Rennie Short.

between vegetation types and altitude. The diagram is marked with the altitudinal limits of plant growth (see Figure 3.2). The atlas is part resource inventory to inform foreign investors, part political project to represent the fledgling Republic as an organic whole and part manifestation of a much wider scientific discourse of how to understand and represent the world.

Mexico

Mexico had major territorial ruptures. The country obtained independence in 1821 as part of a much larger entity until the separate republics of Central America split off in 1823. In 1824, it annexed what is now the state of Chiapas and later, in 1848, most of the Yucatan. In the north there were boundary disputes with its expansionist neighbor, the United States. In 1836, Texas, with encouragement of the United States, declared independence. Border disputes erupted into the U.S.–Mexican War of 1846–48. It was a major victory for the United States but a disaster for Mexico. Under the 1848 Treaty of Guadalupe Hidalgo, it lost almost half of its national territory, including land that includes the present states of Arizona, California, Nevada, New Mexico and Utah as well as parts of what are now Colorado, Oklahoma and Wyoming. In this fluid situation, the Mexican state sought to replace vulnerable spaces and places of ambiguous meaning by fixing boundaries and locating them in administrative territorial structures.[5]

The national atlases were means to fix the state in time and space at a time of territorial loss, border insecurity and lack of national cohesion. Antonio Garcia Cubas, an important geographer in Mexico, produced an atlas in 1858, *Atlas geográfico, estadístico, e histórico de la república Mexicana*, that contained the first printed map of Mexico after the huge territorial loss to the United States. It depicts a diminished national state with fixed international borders.[6] Later, in 1885, he produced the *Atlas pintoresco é histórico de los Estados Unidos Mexicanos* under the enlightened dictatorship of Porfirio Diaz, and after Mexico had regained some confidence after the territorial loss. His map of Mexico was shown at the 1889 World's Fair in Paris: it shows a country of internal spatial coherence and globally connected via shipping lanes and telegraph lines.

The *Atlas pintoresco é histórico de los Estados Unidos Mexicanos* was published in 1885 and produced to be displayed at the 1889 World's Fair held in Paris. It is a multimedia text with a wide range of images, including maps, lithographs and diagrams in 13 plates (see Figure 3.3). The modernizing of the country through better transport links is a recurring theme with many images of new railway bridges and the depiction of the tracks of shipping routes. The atlas is a multimedia, multilayered text with maps surrounded and encased

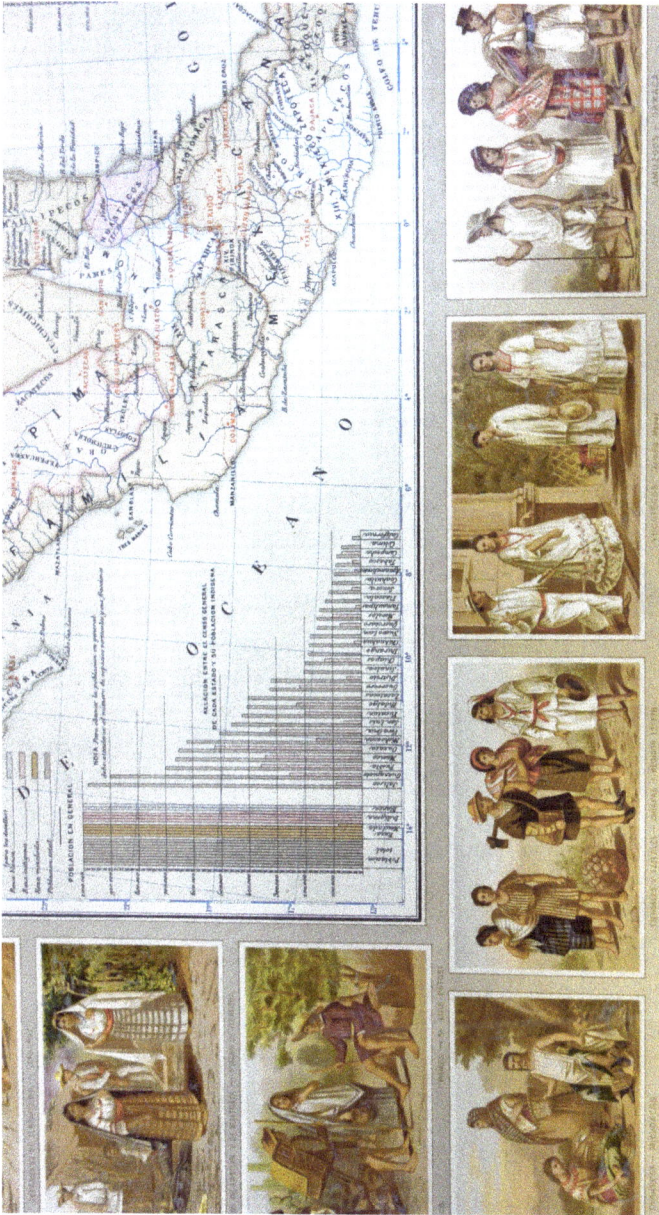

Figure 3.3 Rich imagery, *Atlas Pintaresco e Historico de Los estados unidos Mexicanos*, 1885.
Source: Photo by John Rennie Short.

in images from lithographs, pictures and paintings. It is a wide-ranging visual presentation of the nation. It displays the country as ripe for investment with images of a spatially coherent, well-connected and resource-rich country. But the images operate more than just as attractive postcards to assure foreign investors. It draws upon rich reservoirs of domestic and foreign imageries of Mexico, its buildings and peoples, ruins and vibrant cities to present a country with a deep history, suggestive of stability and legitimacy not a fragile, postcolonial upstart. The images "unify and frame the fragments of real and imagined Mexico."[7] It is not innocent imagery. It is imbricated with the elite Mexican notions of class, gender, race and ethnic hierarchies. The statistical diagram in Figure 3.3 classifies the population into three racial categories of white, indigenous and mixed race—a topic we will discuss in more detail in Chapter 7.

From Anxieties to Certainties

The modern national atlas produced at a time of cartographic anxiety sought to invent a national space, establish a national identity and create a national community. It created territorial order from postcolonial chaos and turned spatial uncertainty into spatial coherence. The different parts of the recently emerged national space were transformed into a national coherence as disparate regions were presented and portrayed as part of a national whole. The 1885 atlas of Mexico, for example, has in its title the evocative, *los Estados Unidos Mexicanos*. The national territory is rendered coherent, fixed, legible and exploitable, binding it to legitimizing geographic and historical narratives. The national atlas created nationally connected places from uncertain spaces, transforming the country into an ordered governed space that is also part of a wider world of global discourses and flows.

Remaining Anxieties

Cartographic anxieties remain even in some of the later more modern national atlases. The 1979 *Atlas de Venezuela*, for example, still makes claim to part of the territory that was described in the much earlier 1840 atlas as "usurped by the English." The 1979 *Atlas* makes a claim to most of Western Guyana up the Essequibo River with a notation describing the territory as a contested area (see Figure 3.4).

Sometimes the territorial anxieties are elided. The *National Atlas of China* in its various editions and forms always portrays Taiwan as an integral part of China. The 1988 *National Atlas of Ethiopia* depicts Eritrea simply as another province of Ethiopia. It is presented as integral part of the state and yet

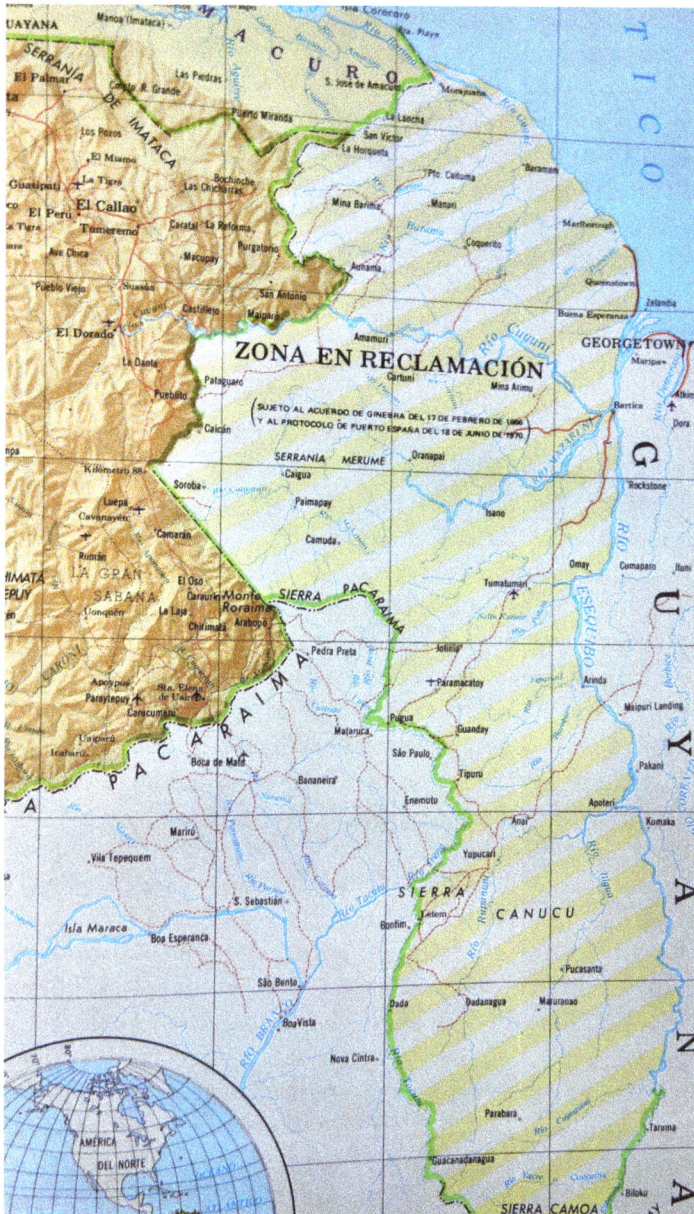

Figure 3.4 Territorial claims, *Atlas de Venezuela*, 1979. *Source:* Photo by John Rennie Short.

Eritrea was only formerly annexed in 1961 and the atlas was produced during the fierce independence struggle fought by the Eritrean Liberation Front from 1961 to 1992. Eritrea's formal independence came in 1993. The territorial breakaway was simply ignored in the atlas.

Some national atlases still have to deal with the shifting contexts of state boundaries and with their legality in international standing. The most obvious case is that of Israel. The 1956 *Atlas Yiśra'el* (*Atlas of Israel*) depicts the state as defined by its early formulation in 1948. In a subsequent edition, the 1970 *Atlas of Israel*, produced after the Six-Day War of 1967 and Israel's annexation of the West Bank and Gaza Strip, the contested nature of the newly annexed territory is described thus: "boundaries of the area mapped presented a special problems [...] coverage now extends in many cases over the entire area under Israeli administration since 1967."[8] By the time of the 1985 edition of the *Atlas of Israel* and the 2011 *New Atlas of Israel*, the annexed territory is depicted as an integral and hence legitimate part of Israeli national territory despite the lack of international approval.

National atlases try to replace cartographic anxiety with cartographic certainty. But even that can prove difficult.

The national atlases of India and Pakistan must deal with the disputed territory of Kashmir. In the 1959, 1982 and 2003 editions of the *National Atlas of India*, all of Kashmir is depicted as Indian territory. The 1972 Line of Control that divides the region into Pakistan- and Indian-controlled territories into a de facto border is simply ignored. The fact that Pakistan-administered area often does not have the data for thematic maps is simply explained as "Data not Available," the implication being that it is Indian territory—it just lacks data. Pakistani mapmakers, in contrast, adopt a different solution. The 1997 *Atlas of Pakistan* describes all of Kashmir as "Disputed Territory" with notations such as "The accession of Jammu and Kashmir to Pakistan and India remains to be decided." All the maps in the atlas show Jammu and Kashmir as "Disputed Territory" (see Figure 3.5). For India, the national atlas portrays a settled land under Indian ownership while the Pakistan national atlas depicts the dispute as a festering, unresolved issue.

When the 2009 *National Atlas of Korea* depicts the national territory, it shows a unified space, North and South as part of one coherent, unified "Korea." The thematic maps however bow to the fact that there is no data north of the Armistice line. So, maps of surface water quality, for example, show the variation throughout South Korea while North Korea is forlornly shown as blank. From rice production to years of schooling, the South is shown in all its variability while the blank surface of a truncated North Korea reinforces the

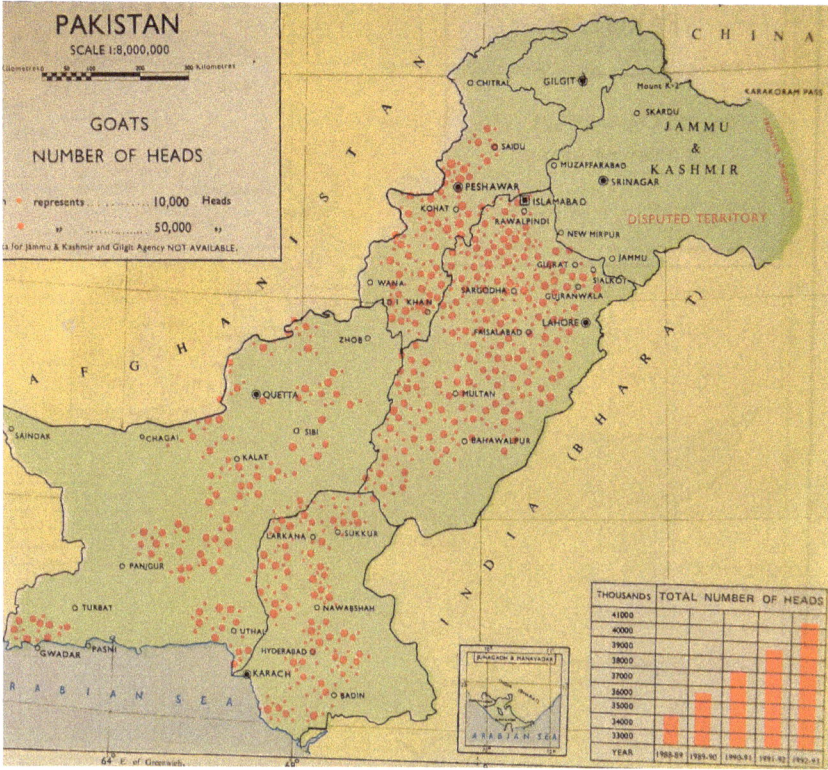

Figure 3.5 Map of goats, *Atlas of Pakistan*, 1997. *Source:* Photo by John Rennie Short.

division. The blankness of North Korea in the thematic maps is fitting symbolism for repressive authoritarianism. The territory is deprived of color and vitality just as the people of North Korea are denied democracy and freedom. The difference between a highly textured South Korea and uniformly blank North Korea is a reminder of the scar that still separates Korea.

Chapter 4

CARTOGRAPHIC RUPTURES AND THE NATIONAL ATLAS

The national atlas often emerged after revolutionary ruptures, imperial collapses and successful colonial struggles, playing an important part after political disjuncture and social disruption, giving shape and form (quite literally!) to new nation-states. The national atlas gave expression to emergent national identities, helped create/recreate national communities, recombined linkages worldwide and expressed changed relations to an ever-widening world.

Cartographic Declarations of Independence

The usual unfolding is when a nation-state emerges, then that new state organizes its cartographic expression, very often in the form of a national atlas. In some cases, however, the national atlas predates the emergence of the new nation-state. Let's consider two examples of a national atlas as a declaration of independence and the national atlas as a "premature birth certificate."

Finland

An early example is the first *Atlas öfver Finland (Atlas of Finland)*, printed in 1899, published in Finnish and Swedish (see Figure 1.1). This atlas has legitimate claims to be the first modern national atlas in the world. A second edition was published in 1910 in Finnish, Swedish and French (see Figure 1.2). At the time of both editions, Finland was not yet an independent country but under the rule of the Russian tsar. Previously, it was under the control of the Swedish Empire. From the thirteenth century to 1809, Finland came under Swedish domination, and Swedish became the language of the Finnish elite. The language of higher education also was conducted in Swedish, even after Finland became part of the Russian Empire.

After the 1808–9 Finnish war between Sweden and Russia, most of Finland was ceded to the Russian Empire as the semiautonomous Grand

Duchy of Finland. This was the predecessor state of modern Finland. Throughout much of the nineteenth century, Finnish cultural expression blossomed, built on its unique language, the Lutheran Church and its own folkloric traditions. Finland was part of the Tsarist Empire but a very distinctive part. Finnish cultural expression strengthened during periods of greater autonomy from Russian control. Finnish, spoken by the peasantry, became a vehicle for nationalism. It was promoted by the Swedish-speaking elite to forge a link of national identity with the masses and by the Russian authorities to separate Finns from the local Swedish-speaking elite and from the neighboring Swedish state. The Grand Duchy was part of the Russian Empire but with more autonomy than other parts of the empire. Overt Russian control waxed and waned, and during bouts of "Russification," Finnish nationalism was inflamed. Russification included the right of St. Petersburg to overrule local assemblies, making Russian the language of administration, incorporation of the Finnish Army and conscripting Finns into the Russian imperial forces. Russification occurred from 1899 to 1917.

With enough relative autonomy in cultural, intellectual and educational affairs to foster nationalist sentiment, Finnish nationalism grew despite and sometimes because of bouts of Russification. The national atlas was an integral part and textual expression of Finnish nationalism. The first edition of the atlas in 1899 was published just as a renewed policy of Russification sought to smother Finnish nationalism. The national atlas then is a product of nationalist sentiment even before the creation of the nation. Finland only achieved independence in 1917, when the Russian Empire fell. But it was only with its third edition in 1925 that it was a "national atlas"—the publication of an independent country. This third edition was published in English, Finnish and Swedish and compared to the first edition, reflected a country moving from rural to urban and from agricultural to industrial. Figure 4.1, for example, depicts the 1920 industrial population.

The *Atlas öfver Finland* was a self-consciously national atlas produced at a time when Finland was not an independent nation and when an active policy of Russification was being pursued by the imperial authorities. The atlas proclaims Finland as a de facto state. The joint publication in Swedish and Finnish represents the cultural legacy of Swedish in the Finnish elite as well as the emerging nationalist sentiments of the majority of Finns. The joint publication also allowed a wider readership for the atlas since Finnish is a little-known language. The 1910 *Atlas öfver Finland* was published in French as well as Finnish and Swedish, giving the atlas even more visibility since French was a major language of the wider European intellectual world, spoken by elites in Poland and Russia and throughout Europe.

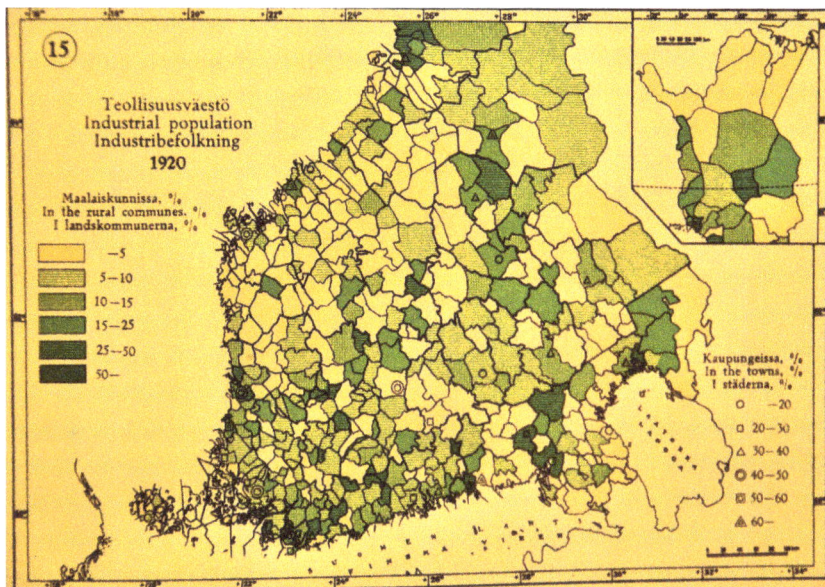

Figure 4.1 Map of industrial population, *Atlas of Finland*, 1925. *Source:* Photo by John Rennie Short.

An atlas of Finland was published on a regular basis. In subsequent chapters, I will draw on the later editions of 1925, 1960 and 1976–95 as well as the first early editions to highlight the changing form and subject matter of the national atlas. For the moment, however, we can note that the first two editions of this highly influential modern atlas emerged when Finland was part of the Russian Empire as a form of national expression and resistance against Russification and the possible obliteration of Finnish identity. The *Atlas öfver Finland* was a way to represent Finland as an independent nation. It expressed not the legal fact of nationhood but the yearning for nationhood.[1]

Poland

For most of the nineteenth century there was no independent Poland. It did not exist as a sovereign state because from 1795 until 1918 Poland was partitioned by the land empires of continental Europe. It was divided up between three imperial powers: Prussia (Germany), Russia and the Habsburg Monarchy. The idea of an independent Poland persisted despite, and also because of, the imperial control. An uprising in 1830 of Poland and another in 1863 pursued the idea of independence. Compared to its imperial overlords, Poland had

a distinct language, the Catholic religion and a vibrant intellectual life that
continued to express the idea of a separate nation.

World War I provided the Poles with an opportunity for independence
when the power of their three imperial overlords was shattered. At the
start of the conflict, "Polish" territory was divided between the Central
Powers of Germany and Austria on the one hand and Russia, in alliance
with UK and France, on the other. Military victories against Russian forces
allowed the Central Powers to proclaim a Kingdom of Poland in 1916 in
formerly Russian-controlled lands. Later, in 1917, the Russian Revolution
of 1917 toppled the Tsarist regime, and the 1918 defeat of the German and
Habsburg Empires inaugurated their postimperial collapse. An independent
Poland became part of a new postwar order. President Woodrow Wilson's
Fourteen Points enunciated in January 1918 to the U.S. Congress specifically
mentioned Poland:

> An independent Polish state should be erected which should include the
> territories inhabited by indisputably Polish populations, which should
> be assured a free and secure access to the sea, and whose political and
> economic independence and territorial integrity should be guaranteed
> by international covenant.[2]

Polish independence was achieved in 1918 and codified the following year at
the Paris Peace Conference and in the Treaty of Versailles. The Paris conference endorsed a new European order that gave birth to a Polish state. It
was an uneasy start for the new state. There were six border wars between
1918 and 1921, the fully fledged Polish-Ukrainian War of 1918–19 and the
Polish-Soviet War of 1919–21. The new nation had to struggle to assert its
territorial right to exist.

One important element in the birth of the nation was the role played by
Eugene Romer (1871–1954) and his atlas of Poland. Romer was an influential figure in Polish cartography. A Polish nationalist, he was also part of
a network of international intellectuals who exchanged letters and ideas.
Romer was well placed. He spoke fluent German and French and was part of
what Steven Seegal refers to as the mapmen of Eastern Europe who were in
constant communication with each other as maps were drawn and atlases
produced.[3] Many of these maps laid the spatial foundation of what is now
Poland, Lithuania and Ukraine.

Romer grew up in the German-controlled part of Poland and studied in
Germany where he came under the influence of German geographical and
geopolitical scholarship. In 1908, he published a geographical atlas of Poland
with an accompanying text. It was concerned mainly with physical geography.

By 1911, he held a chair in Lwow, then part of the Austro-Hungarian Empire. At that time Social Darwinism was influencing geopolitics. Nations were imagined less as political constructs and more as "natural" biological units that needed to grow or die, locked into a never-ending struggle for the domination of space. It was the "survival of the fittest" applied to international relations. This Darwinian imagining of the nation-state became an important theme of German political geographers. Friedrich Ratzel (1844–1904) argued that a rapidly growing Germany needed more space, *lebensraum* (living space), in order to grow.[4] States were locked into a competition for territory. This was the intellectual backdrop for Romer's idea of Polish territorial individuality and his cartographic claims for a Polish state.

At the beginning of World War I, Russian forces captured Lwow, and Romer with his family fled to and settled in Vienna. In exile, he envisioned an atlas to justify Polish demands for independence at what he anticipated would be an international congress at the end of the war. Romer was at the center of a group of Polish exiles looking for ways to achieve an independent Poland from the flux of the war. After Central Powers took Lwow in 1915, he returned, and in 1916 published *Geograficzno-statystyczny atlas Polski* (*Geographical and Statistical Atlas of Poland*).[5] It was published in Warsaw and Cracow. What was remarkable was that Romer used the term Poland to cover 800,000 square kilometers that then belonged to Austria, Germany and Russia (see Figure 4.2). It was a very expansive interpretation of Polish lands. Romer's Poland included most of present-day Poland as well as large chunks of present-day Belarus, Lithuania and western Ukraine. He collected and represented official data from the three separate states as one spatial unit, thus erasing the existing boundaries of the three empires to cover territory that Romer argued was ethnically Polish. As one modern commentator noted, "Poland was revived in the cartographic space. His method of plotting the data to erase the outlines of the three states involved was a clever way of uniting Polish lands."[6]

The atlas contained 32 plates and 70 maps covering the physical geography, history and the demographic and economic aspects of 441 separate territorial units. Each map had a brief commentary. The atlas was published in German (then considered the language of science), French (then considered the language of diplomacy) and Polish. Polish names were used in the maps. It was "Gdansk" not "Danzig," for example. A review by German geographer Albrecht Penck in 1917 criticized the use of French as the language of the enemy and the expansive vastness of "Polish lands." The atlas was smuggled to the West and displayed at international conferences in 1916 and 1917; it was cited at the 1919 Paris Peace Conference. In discussions of Poland's borders, Romer's maps were cited as opposed to German and Russian claims

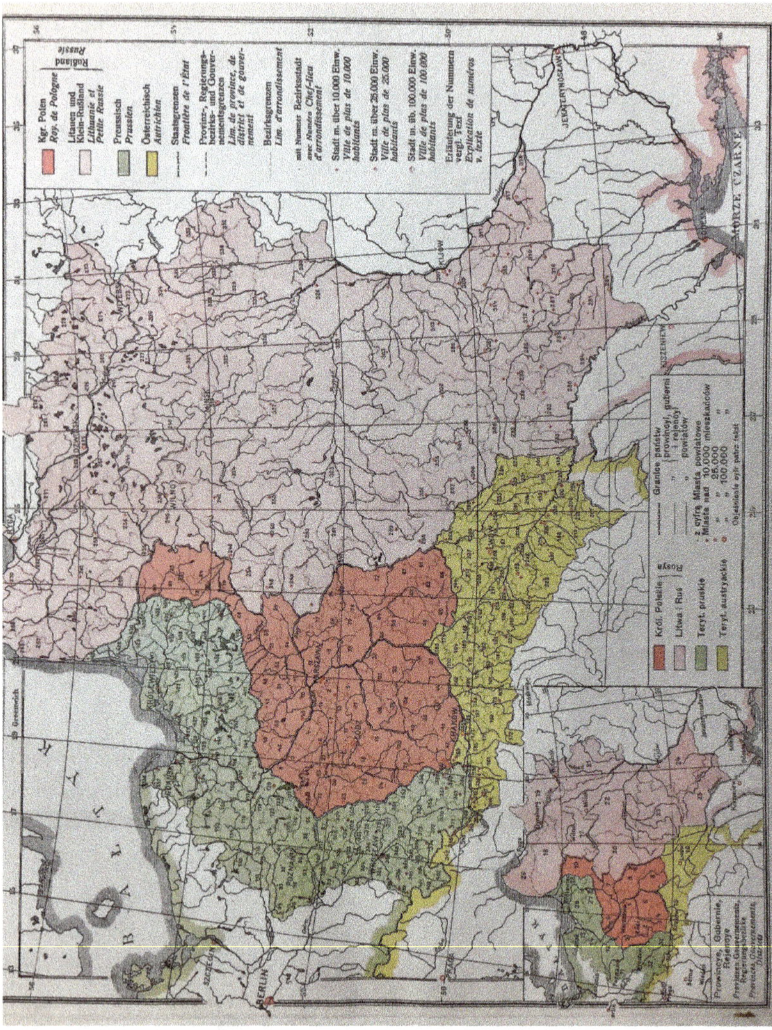

Figure 4.2 Map, *Atlas Polski*, 1916. *Source:* Photo by John Rennie Short.

and as against British reluctance. But the maps actually used rather than merely cited were not from his atlas but were new maps, later gathered in *Atlas Polski Atlas Kongresowy (Congress Atlas of Poland)* and published in 1921. It was the spirit of Romer's atlas rather than his actual maps that were used at the 1919 Peace Conference.

Romer went on to play an important role in an independent Poland. He encouraged national cartographic endeavors and promoted the international scientific ties of the new nation. He helped, for example, to organize the International Geographical Conference in Warsaw in 1934. He died in Krakow in 1954. His greatest achievement was to create an atlas that helped to give birth to an independent Poland. The state that he imagined and represented in his 1916 atlas bore a striking resemblance to the Polish nation-state that emerged in 1918–19. His aspirational national atlas was a cartographic declaration of independence, the "birth certificate of a nation."[7]

Revolutionary Ruptures

The national atlas also emerges from revolutionary ruptures. The more dramatic the rupture, the greater the perceived need to reimagine the nation-state in order to establish a new cartographic identity as well as the need to educate both the national and international communities of the reality of a new state.

The Great Soviet World Atlas

Arguably the greatest political rupture of the twentieth century was the Russian Revolution of 1917 that inaugurated not only a new state from the former Tsarist Empire but also a new world order that, until 1989, challenged the hegemony of the West. The establishment of the Soviet Union was an epoch-making event. Such a rupture initiated and prompted new cultural expressions. The new Soviet Union wanted to heighten the sense of national order and cohesion. A new cultural language emerged involving slogans, posters, festivals, visual art, film, theater and literature. In the earliest years, there was a vibrant avant-garde aesthetics that reveled in the sense of a "new order." Later, a Stalinist-inspired Soviet Realism deadened this cultural vitality.

Images played an especially important role in propaganda purposes for an illiterate or poorly schooled population. The early Soviets created new forms of cultural expression, and maps were an important part of the ensemble of this vibrant propaganda art and culture. The artist Lazar Markovich Lissitzky (perhaps better known as El Lissitzky), a follower of Kazimir

Malevich, produced innovative work that influenced the Bauhaus movement. His 1919 poster/map, *Beat the Whites with the Red Wedge*, produced during the Civil War (1918–21) is a stylized geometric map/diagram that shows a triangle of red pushing against and puncturing black forces spread across the territory of the Soviet Union. Maps that combined Russian avant-garde art with traditional Russian icons and folk art also were produced. Printed in 1928, a set of 10 propaganda maps was produced as a series of posters by the Division of Military Literature of the State Publishing House. These maps combine traditional cartography with figurative images of the major events of the revolution and the Civil War. The third map in the series, *Plan of the Entente to Suffocate the Soviet Regime, May–October 1918*, depicts a small red area around Moscow surrounded by the troops and forces of the alliance represented by flags of Czechoslovakia, France, Great Britain, Imperial Russia and the Provisional Government of Siberia. American and British battleships are depicted steaming toward the Russian Arctic coast. These large maps were meant for display in public places such as schools, offices and factories. Together they recount the emergence of the Soviet Union and its vulnerability to outside interference and ultimately its secure success. Cartography was a propaganda tool and maps were an explicit form of political rhetoric.

Tsarist Russia has a long cartographic tradition of mapmaking and map publishing. The sprawling empire was mapped and surveyed as it was incorporated and controlled. But there was no formalized national atlas. As early as 1921, Lenin suggested the need for a national atlas for the new state. The *Bol'shoĭ sovetskiĭ atlas mira (Great Soviet World Atlas)* was a massive undertaking. It began in 1934 under the direction of V. E. Motylev and was published in two volumes and only in Russian. It is an atlas that celebrates the industrialization, modernization and collectivization of the USSR. Regional maps of the country take an economic inventory—especially of electricity production— reminding us of the famous quote attributed to Lenin that "Communism is Soviet power plus the electrification of the whole country."[8] The atlas also depicts the collectivization of agriculture. There are maps of tractor stations, collectivized farms and new railroad stations. The maps display a modernizing and collectivizing economy. The second volume consists of regional maps with an emphasis on economic geography. The *Great Soviet Atlas* is an economic inventory that emphasizes rapid industrialization and socialist economic growth.

The first volume was published in 1937. The title page includes the embossed heads of Lenin and Stalin and the quote, "Workers of The World, Unite!" It contains 83 thematic maps of the world and 36 thematic maps of the USSR. The world thematic maps include production levels for aluminum, pig iron and steel production as well as machine construction and electrification. It is a global economic inventory that reflects the Soviet rapid industrialization and inadvertently the insecurity of its standing with major capitalist powers. Cartographic depictions of the whole country are divided in two with a European and an Asiatic USSR. It is a way to deal with the vastness of the territory. But it also reveals the nature of Soviet control and geographical imagination. Allegiance to the Soviet state was strongest in the Russian industrial and urban centers of the West. And it is also perhaps indicative of an old trope of a West-East divide between what was considered a more modern, more European West compared to the backward Asiatic territories of the East.[9] There are also a series of maps that describe global capitalism. Maps with such titles as "Map of exports of imperialist nations," "Map of the raw material market of imperialist nations," and "Map of financial relationship of capitalist countries, the export of capital" (see Figure 4.3) signify a new understanding of the geopolitics and geoeconomics of the global capitalist economy that surrounds and threatens the USSR.

The second part of Volume 1 is devoted to the USSR. The concerns of the regime are highlighted. There are maps that constitute an economic resource inventory and measures of modernization. Improvements in electrification are highlighted because they meant power in the double sense, the ability to modernize and industrialize rapidly and also a demonstration

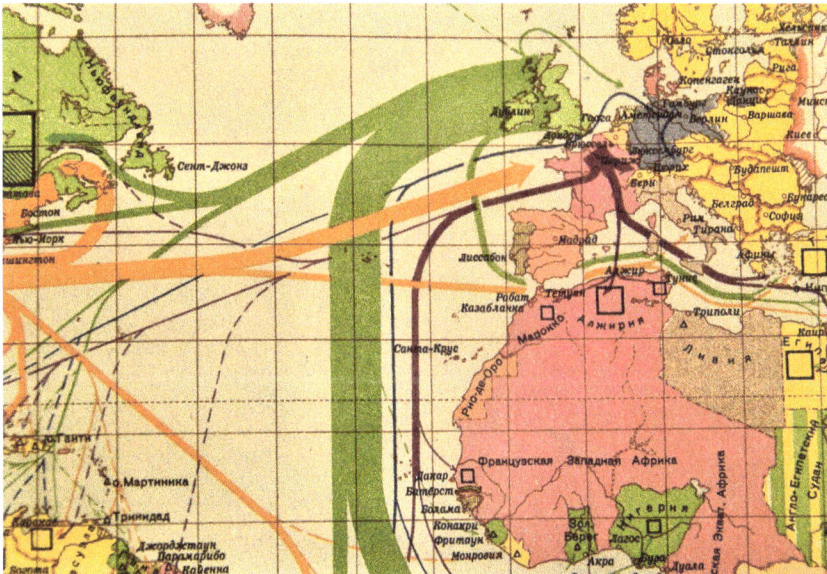

Figure 4.3 "Map of financial relationship of capitalist countries, the export of capital," *Bolshoi sovietski atlas mira* (*Great Soviet World Atlas*) 1937. *Source:* Photo by John Rennie Short.

of the state's ability to affect change and improvement. Electrification was a major focus of the Soviet regime since its inception. The atlas reflects this enduring concern. But the atlas also embodied the shift toward more specifically Stalinist concerns. In 1921, a New Economic Policy (NEP) was introduced to cope with food shortages and mounting social unrest. It was a policy that involved mingling of the free market with Soviet control; it allowed peasants to sell food on the open market, fostered the denationalization of small-scale industries and the encouragement of foreign investment. The NEP involved a return to more market forces after the war. The NEP was successful, and by 1927 the economy had recovered to prewar levels, which was no mean achievement after the devastation of World War I and the ensuing Civil War. But the policy also saw the emergence of a capitalist class of kulaks (richer farmers and landowners) and so-called NEPmen (small business owners) who were making money in the private market in the cities. When Stalin gained control in the mid-1920s and ultimately achieved absolute power, he reversed course. He pushed for the collectivization of agriculture and rapid industrialization. He introduced central planning (the first five-year plan took effect in 1932) and renationalized most of the economy. The post-NEP Stalinist policy is apparent in the atlas with its emphasis on collectivization. Graphs depict the yearly rise in the number of collective farms, and maps for 1928, 1930, 1933 and 1936 show a "creeping" collectivization, shown in red across the national territory. Figure 4.4 shows the level of collectivization in 1936 with the graphs in the bottom

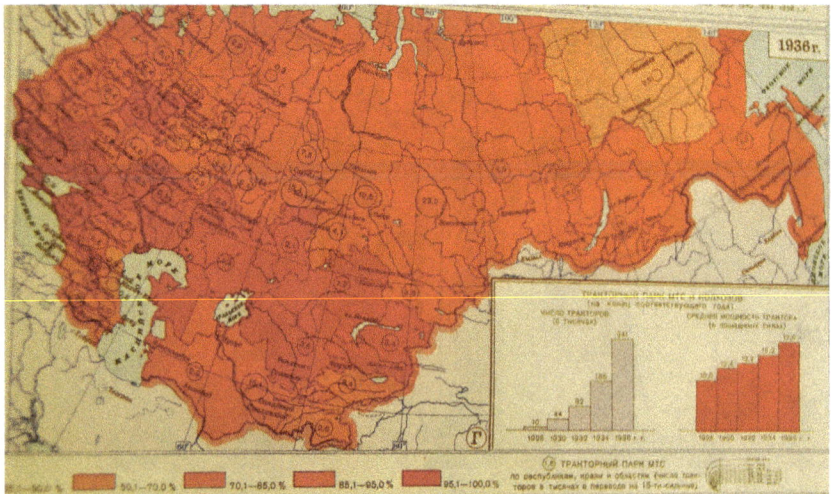

Figure 4.4 Collectivization of farming, *Bolshoi sovietski atlas mira* (*Great Soviet World Atlas*) 1937. *Source:* Photo by John Rennie Short.

right depicting the growing number of tractors and their average horse-power, a sure sign of the mechanization of the collective. The number of collective farms increased from 10 percent of all farms in 1928 to 90 percent by 1936. There are maps of machine tractor stations: what the maps do not show—and all maps have silences as well as utterances—are some of the costs of this policy shift. The Soviet famine of 1932–33, in large part the result of forced collectivization and forced grain procurement, killed millions, especially in the main grain-producing areas in Ukraine, the Volga region and Kazakhstan. Close to 10 million died from starvation across the Soviet grain belt.

Volume 2 was published in 1938–39. It includes detailed economic maps of different regions and republics of the USSR. There is some interesting cartography, but most maps are too crowded as they try to depict too many detailed branches of industry, electric stations, deposits of useful material and transportation. City maps point out electrified rail lines and new subway status. The maps are a report of modern industrializing cities. Volume 2 also contains maps of the course of the recent Civil War. The *Great Soviet Atlas* records and celebrates the socialist revolution.

With growing hostilities and the fear that information in the atlas could become part of enemies' knowledge of the territory, the distribution of Volume 2 was halted soon after it was produced. It was considered a security risk because it contained strategically important information about the distribution of national assets, at a time of war. Few copies of the atlas remain.

The *Great Soviet Atlas* is a major accomplishment that because of World War II and its publication only in Russian did not receive much international attention. It was produced at the height of Stalin's power and reveals the enduring concerns with electrification and Stalin's emphasis on industrialization and collectivization and his growing political paranoia. Maps of industry compared the Russia of 1915 with the USSR of 1935 to highlight the rapid industrialization. It also revealed the insecurities of the state in the social categories used to count and distinguish the population. The maps employed a population categorization of

workers and employees,
kolkhoz (collectivized) peasantry,
individual peasant farmers,
bourgeoisie,
other population (students, pensioners, army and others).

The bourgeoisie included landowners, large and small bourgeois, tradesmen and kulaks (peasants with over eight acres of land who employed nonfamily labor). This was not an innocent counting. Stalin ordered the kulaks to be liquidated. Estimates of total deaths vary from 700,000 to several million. Former kulaks made up the majority of victims of the purges of the late 1930s with almost a million either arrested and sent to gulags or executed.

New socialist states

The Soviets liberated most of Eastern Europe from German occupation. Soon with the help of national communist parties they undermined democratic parties and took political control in what was then described as Eastern Europe. As the countries of Eastern Europe became socialist states, new atlases appeared, including Czechoslovakia (1966), Hungary (1967 and 1989), Bulgaria (1973), Romania (1979) and German Democratic Republic (1981).

The national atlases of these new socialist states make for interesting reading. Most of them were published in part of a propaganda exercise for internal and external consumption celebrating modernizing, industrializing and socialist societies. The atlases were also an integral part of national planning; the atlases were often used to take national inventories, complement national economic planning and inform spatial planning.

Romania

The *Atlas Republica Socialistă România* was published in 1974 by the National Academy and Institute of Geography. This was the second atlas published by the Romanian regime. The first was published in 1965. The first maps in the 1974 edition highlight Romania in a map of the world. There are standard maps of physical and social phenomena, including traffic flows through Bucharest (Figure 4.5), which is the most distinct map in an otherwise more traditional repertoire of cartographic depictions. The map looks like a set of arteries coursing through the heart of the city. The atlas was published in Romanian, English, French and Russian. It is a very large volume with maps of double page folio. The size reminds us of the socialist gigantism of Romanian state socialism. Official public buildings, like the atlas, were made so large to make the individual citizen so small. The introduction notes that Romania has lost its agrarian character and is now an industrial-agrarian country. What is not noted, for obvious reasons, was the catastrophic shock therapy policies of the leader Nicolae Ceausescu. To boost the population, he prohibited abortions—then freely available in Eastern Europe and the USSR—for women under forty with less than four children. The results were high infant mortality rates, child abandonment and ultimately more

Figure 4.5 Traffic flows in Bucharest, *Atlas Republica Socialista Romania*, 1974. *Source:* Photo by John Rennie Short.

than 100,000 children institutionalized. The traumatized state of Romanian orphans left uncared for in state institutions was a searing image of the Ceausescu legacy. Economic policies were aimed less at public welfare and more at paying off foreign debt by squeezing the population with high prices, rationing of essentials and forced labor days. The agrarian transformation hailed in the introduction of the atlas included the destruction of over 7,000 villages while the touted modernization included the razing of 40,000 dwellings in Bucharest. So, knowing all of that, it is ironic perhaps to read in the introduction:

> The country continues to build a comprehensive socialist society and strives thereby towards summits of progress and civilization.[10]

It is also important to note that Ceausescu's Romania received favorable treatment from the West; he was considered one of Europe's good communists. His appalling treatment of his own population was politely ignored. He was given the British seal of approval with an official visit to meet the Queen and praised by the United States and Western European powers because he critiqued the Soviet Union and paid back foreign loans.

Soviet influence extended beyond Eastern Europe. Communist Party ideology informed various revolutionary movements including in Afghanistan, Ethiopia, Cuba. Let's look at each of these in turn.

Afghanistan

The USSR was the biggest aid donor to Afghanistan even before the 1978 takeover. It had long been in its geopolitical sphere of influence ever since the days of the Great Game with the UK.

The Democratic Republic of Afghanistan was proclaimed in 1978. A self-proclaimed Marxist–Leninist regime was established, with close ties to the USSR, that sought to establish civil rather than a religiously ordered society. Beards were discouraged. Women's rights were improved. At the same time, the government launched repressive measures against its opponents with killings and imprisonments. Protest mounted and to prop up the collapsing regime, Soviet troops entered the country, killed the unpopular leader and then had to fight a long campaign against Afghan forces aided by the United States, Saudi Arabia and Pakistan. Eventually, the religious zealots of the Taliban came to power until they too were toppled by the U.S. forces in late 2001.

A national atlas was produced in the early years of the new socialist regime. In 1980, the Polish government, probably due to Russian influence, decided to help produce and subsidize a national atlas. Afghanistan's first national atlas, the *National Atlas of the Democratic Republic of Afghanistan*, was published in 1985 in Warsaw: it was printed in two editions, English and Dari. There is the usual socialist rhetoric in the introduction:

> The country's close economic ties with the USSR helped in the construction of a great majority of leading plants and facilities in all branches of the economy. [...] Following the Revolution, there has been a sharp increase in the growth of industrial production.[11]

The introduction expands the idea of developing educational opportunities and medical health provision. This was vital for one of the poorest countries in the world. The geopolitical situation is also addressed in the recurring theme of constant interference in the country's internal affairs by the United States and some other countries. The rhetoric continues with the description of a

> heroic people building a new life under the leadership of their vanguard, the People's Democratic Party of Afghanistan.[12]

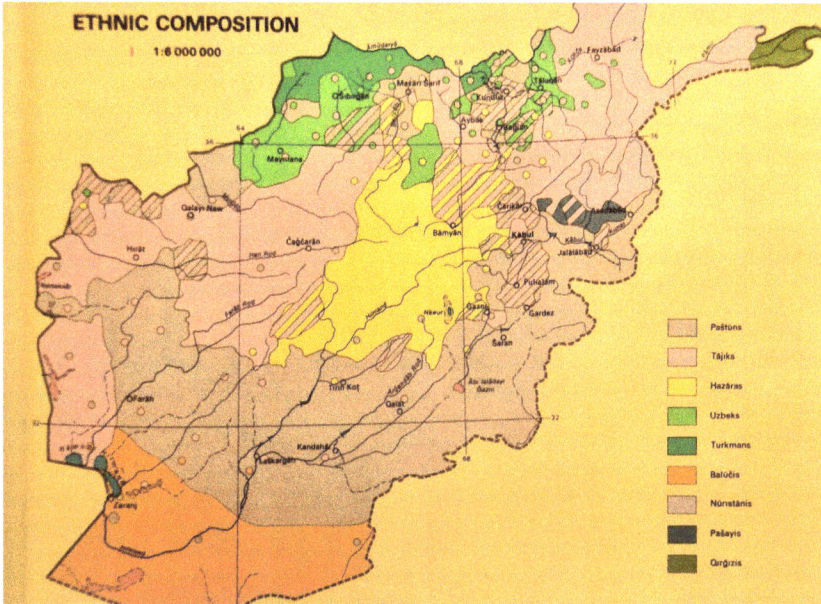

Figure 4.6 Map of ethnic composition, *National Atlas of the Democratic Republic of Afghanistan*, 1985. *Source:* Photo by John Rennie Short.

In comparison to their revolutionary rhetoric, the maps' forms are traditional. The "ethnic composition" map is interesting because of the different categories employed (Figure 4.6). Nine separate groups are distinguished. And the map of tourist sites includes a symbol for the magnificent giant Buddha figures of Bamiyan, once the largest standing Buddhas in the world. Sadly, they no longer exist. They were blown to smithereens in 2001 by the Taliban who saw them as idolatrous remnants of pre-Islamic culture.

Ethiopia

Ethiopia was long ruled by Emperor Haile Selassie who came to power in 1916. Apart from a brief period of Italian occupation, from 1936 to 1941, his rule lasted until the early 1970s when food shortages, high gasoline prices and strikes all undermined his rule. In 1974, a Soviet-backed junta gained control in a coup. A socialist Ethiopia gave the Soviets an ally and possible vital base in east Africa. During the regime's grip on power from 1974 to 1991, the nation's first national atlas was produced. The *National Atlas of Ethiopia* was published in 1988 by the Ethiopian Mapping Authority and printed in Addis Ababa. It was based on work for an earlier national atlas that was published in provisional form in 1982. The production of the atlas was helped by Canadian aid and technical assistance. The atlas was jointly funded by Canada's International Development Research Centre. The text is in English. Published at a time of hardship of war and famine, it contained the standard elements of national atlas. The first maps depict Ethiopia as part of Africa then in the more detailed regional setting of the Horn of Africa. A range of thematic maps cover desertification, locusts and drought, food shortages and major diseases. The standard rhetorical introduction is muted. The prime minister presents the atlas as something that

> draws together the historic moment in our country marked by the foundation of the People's Democratic Republic of Ethiopia and the challenge which now faces our Republic in ensuring the continued advancement of the well-being of our country.[13]

The atlas had the explicit intent of providing a socioeconomic and natural resource of the country. More than seventy-six separate maps cover a wide range of topics including desertification, drought and food shortages. A heart-rending series of maps entitled *Wars and Invasions* depict colonial incursions by British, French, Italian and Ottoman forces as well as what is referred to as the Somali War of Aggression of 1964 and 1977. One map depicts food shortages. The atlas meant to celebrate the revolutionary rupture also records the enormous difficulties of the new nation. Figure 4.7, for example, depicts the country's road transportation network. Notice how wet season accessibility is much less than the dry season accessibility, an indication of the rudimentary infrastructure. The atlas is less a celebration of a new socialist state than an honest portrayal of a country facing major problems of mass starvation, illness, national security and lack of basic infrastructure.

Cuba

Cuba has a range of atlases produced at very different periods in its modern history. First, a very brief recap. Cuba was a Spanish colony that came under the influence of the United States. After Spain's defeat in the Spanish-American War, Spanish troops departed the island in 1898. The next month, January 1899, Cuba came under the direct control of the United States. In less than five years, almost 10 percent of Cuba's land was taken over by American companies and over 80 percent of sugar plantations and cigarette factories were owned by American companies: it was a classic colonial plantation economy. Cuba was a valuable economic colony of the U.S. Empire. This economic imperialism lasted long after the formal political control.

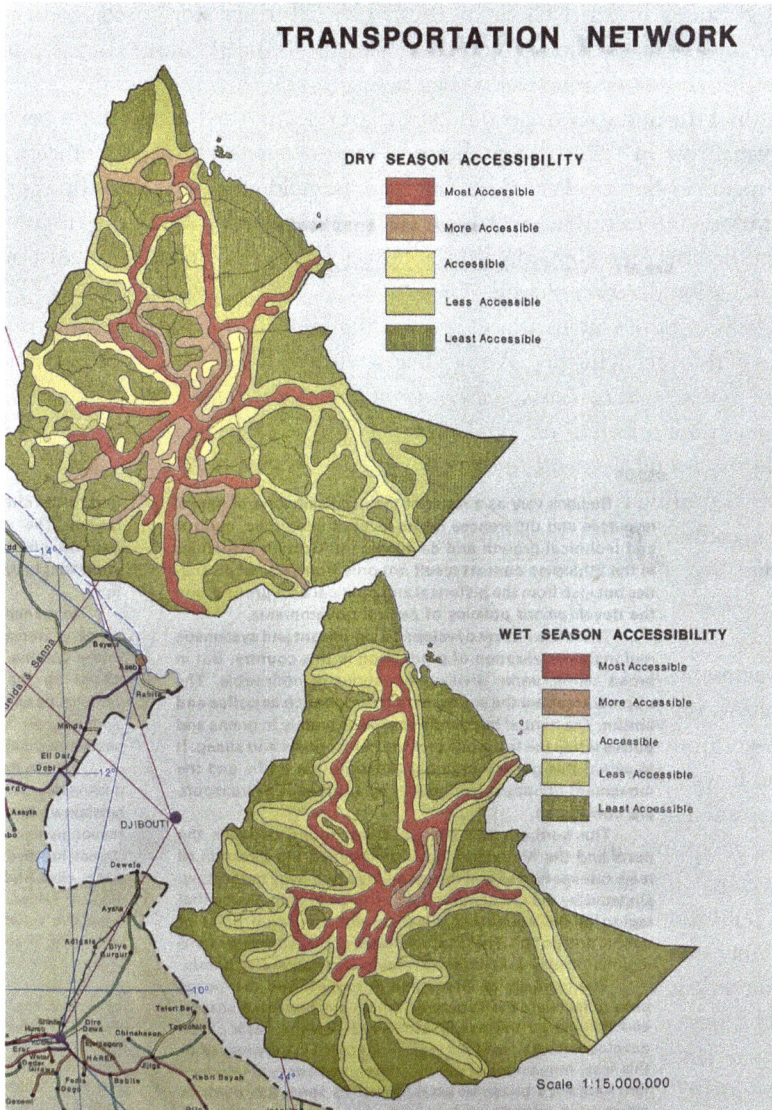

Figure 4.7 Transportation network, *National Atlas of Ethiopia,* 1988. *Source:* Photo by John Rennie Short.

Cuba achieved a limited form of political independence in 1902, but, under the Platt Amendment, U.S. forces would leave yet the United States retained effective control over foreign relations, the size of the national debt and was allowed to intervene in Cuban affairs. Cuban political independence was undercut by economic dependency and effective U.S. political control. The

United States invaded Cuba in 1906. U.S. Marines were used again and again in various invasions/interventions in 1906–9, 1912 and 1917–22 to protect U.S. economic interests. When a more independently minded government had the audacity to propose agrarian reform, the United States backed an overthrow in 1934 by a coalition of landowners and military officers led by Fulgencio Batista. With U.S. backing, he held onto the reins of power as military dictator until his overthrow in 1959 when Fidel Castro rose to power. Cuba then became aligned with the Soviet Union and enacted socialist policies of nationalization of land.

The history of Cuban atlases is revealing as it mirrors the changing trajectories of this recent history. We can begin with an 1884 *Planos de Comunicaciones de Las Provincias de La Isla de Cuba* (*Atlas of Communication of the Provinces of Cuba*). Although not a formal national atlas it is still of interest. Printed in Havana while the island was still under the control of the Spanish, it shows roads, railways, sailing routes and underwater cables. It is the mapping of an island that is connected and modernizing. The map was probably used to stimulate foreign investment; "you can invest in this region" is the underlying message because it is part of a globalizing, connected world. In 1898, *An Atlas of Ports, Cities and Localities of the Island of Cuba* was published by the U.S. War Department. It is a military atlas, a document for an occupying power with emphasis on forts and defenses, approaches and detailed street plans. The U.S.-continued domination of the country is revealed in the 1949 *Atlas de Cuba*. It was published in English and Spanish by Harvard University Press and printed in the United States. The main author is Gerardo Canet, a professor of geography in Havana, but the collaborator is Erwin Raisz of Harvard University. The intellectual neocolonialism is evident in that an atlas of Cuba was published in an American city by an American university press with an American coauthor in a text that at times justifies U.S. intervention as a democratic act to ensure Cuban independence.

The first atlas of the new Castro regime was the *Atlas nacional de Cuba* (*National Atlas of Cuba*). Work began in 1965, completed in 1968 and published in 1970 to celebrate the 10th anniversary of the revolution. The USSR Academy of Science was involved actively, and a Russian-language version was published simultaneously in 1970. Another edition was published in 1978 with more photographs. The introduction to the 1970 edition notes that the atlas should record

the successes in the revolutionary struggle for freedom, the construction of socialism, the threat of American intervention, the establishment of absolute independence and sovereignty of the Cuban states and the transformation of its economic life.[14]

This national atlas presents a new history. The U.S. intervention, unlike the 1949 atlas produced in the United States, is now presented as a military occupation and act of economic imperialism. Historical maps in the 1978 edition show the impacts of "Agresiones del Imperialism." The national atlas is employed not only as a rhetorical device but also as a technical document to record progress and assist state organizations with planning and education. This atlas was made with Soviet aid and assistance. The theme of revolutionary insurrectionary struggles against imperialist aggressions runs through these earlier atlases, but later editions, especially the 1989 edition, lose some of the revolutionary fervor. As the regime became more embedded, there was perhaps less need for legitimation, and the so atlases make a greater claim to science and knowledge than to political ideology. However, the revolution is still celebrated, and the database point of 1959 is used in the maps to reveal the sense of progress and improvement from the dark days of U.S. imperialist control (see Figure 4.8). The national atlas celebrates the revolutionary rupture by noting social and economic progress made since the revolution.

The Postcolonial

The breakup of empires and decolonization in the twentieth century created a plethora of new states. Some of them had a form of historical legitimacy; others were simply the product of postimperial division. The national atlases of the postcolonial countries were used not only to record the new nation but also to create it.

Canada

It may seem odd to include Canada in a section entitled "postcolonial." Canada did not so much throw off British rule as its southern neighbor did, but peacefully passed from colony to dominion to independent state, a progress reflected in respective atlases.

Canada's emergence from its connection with the UK was slow and gradual. Unlike its neighbor to the south, Canada's separation from the United States was neither violent nor sudden, marked more by a peaceful and steady civility than an outright rupture. In the early twentieth century, Canada was independent from the UK under the Constitution Act of 1867 but still part of an imperial system. It was only in 1931, under the Statute of Westminster, that Canada became fully sovereign. This liminal space, somewhere between independence and full sovereignty, is reflected in a comparison of atlases of the time. In 1905, *The Imperial Atlas of the Dominion of Canada and the World* was published by the Arnt-Gill Company in Toronto. In this atlas, the nation

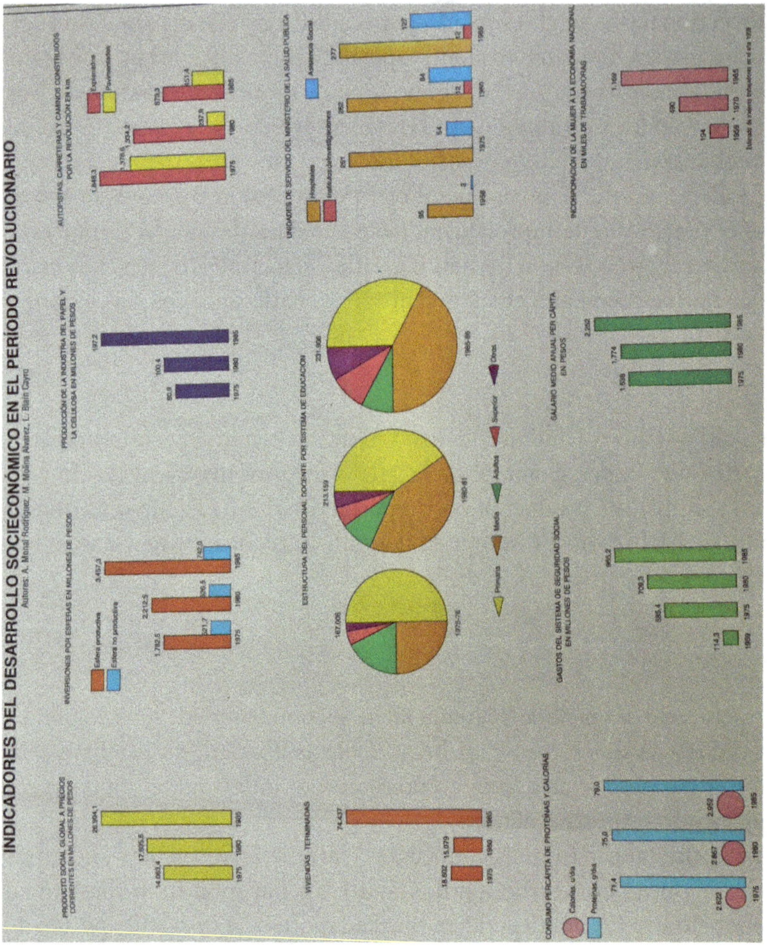

Figure 4.8 "Indicators of socioeconomic development in the revolutionary period," *Nuevo Atlas Nacional de Cuba*, 1989. *Source:* Photo by John Rennie Short.

is framed as part of a wider British Empire. The nation is described as the dominion of Canada, one section is entitled, "Maps of the British Empire throughout the world," and in the very first map, a map of the world, Canada, as with all the other colonies and dominions, is colored red. The nation is celebrated as part of the British Empire and the defeat of a "wilderness" described thus,

> untilled plains, silent and lonely mountain ranges, and untracked forests covered the site of what has now become the most important factor in the mighty British Empire.[15]

There is little room for the indigenous people, and the history depicted is of the colonial struggle between Britain and France. Canada is shown as part of an imperial system, an important member of a global empire.

A year later, the Department of the Interior published the *Atlas of Canada*. While the *Imperial* atlas was a commercial undertaking, no doubt saving costs by recycling maps from other publications with the maps of the empire easily copied from one imperial atlas to the next, this atlas was a distinctly national affair. The aim was less to highlight connections with the British Empire and more to show a separate country. It is a cartographic and statistical compendium of a new nation. The first map shows the country as a singular entity, composed of various territorial divisions. The atlas celebrates progress with maps of telegraphs, telephone links, railways and canals, and the street maps of the major cities. It also shares a colonial bias with the previous atlas; routes of European explorers dominate the historical geography. The atlas is also a celebration of an economic nation with maps of trade, industry and bank deposits. It is a commodified and commercialized nation-state that is shown and celebrated. A revised and enlarged atlas was published in 1916 and again in 1958 with a French edition published in 1960. And as bilingualism became more embedded, the 1974 edition of the *Atlas of Canada* was simultaneously published as *National Atlas of Canada/Atlas National du Canada*. The prime minister's dedication notes that the atlas

> provides a window on the pageant of Canada's progress [...] an achievement made possible, above all, by the unified effort of Canadians acting in the spirit of national purpose.[16]

The dedication can only be fully understood in the context of the time of rising Quebec nationalism and threats of separatism. The dedication was as much a hope for the future as an affirmation of the present. The map in Figure 4.9 is a space-time map that illuminates the "shrinkage" of Canada

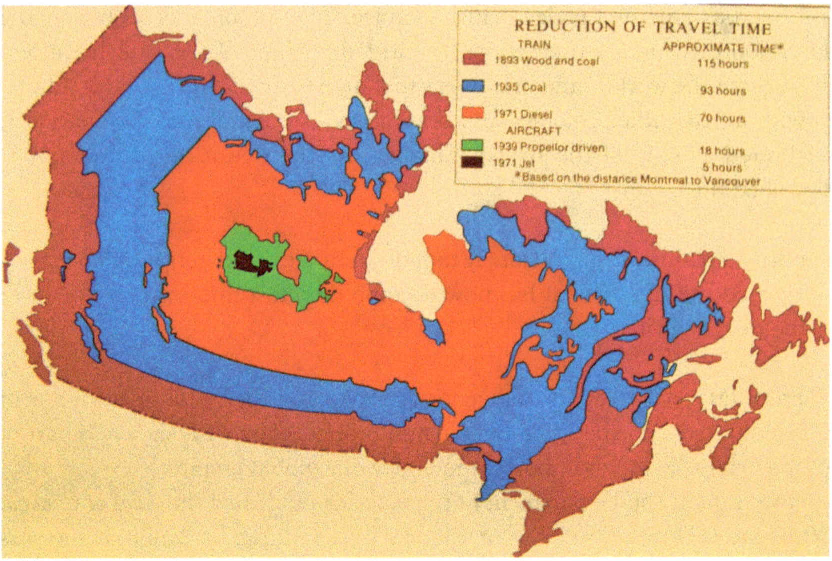

Figure 4.9 Space–time map of a "shrinking" Canada, *National Atlas of Canada*, 1974. *Source:* Photo by John Rennie Short.

in relation to the time traveled. It is suggestive of a country pulled tight into a national embrace. It depicts an integrated, coherent nation, pulling together in space-time closeness.

Pakistan

A national atlas of Pakistan was published in 1985, under the title *Atlas of Pakistan* and a second edition was published in 1997. In 2012, with a nod to the rise of a more explicit Islamic influence in the country and in society, the third edition of the atlas was retitled *Atlas of Islamic Republic of Pakistan*. In the preface to the second edition of 1997, the deputy surveyor general noted that until independence,

> the atlases available in the country till then had all been foreign-drafted, depicting Pakistan insignificantly and portraying foreign political concepts, specifically in matters relating to international boundaries while those few drafted and published by private sector as commercial projects presented none or limited phase of socioeconomic life of the nation.[17]

The "international boundaries" in this case refer to the disputed territory of Kashmir which the atlas depicts in some maps as part of Pakistan national territory while in other maps as disputed territory with an undefined frontier (see Figure 3.5).

India

The first national atlas of an independent India was published in 1959. In a self-conscious break with the British imperial system, the metric system was employed, and Hindi was used in the title, legends and notes, although there were still 25,000 words of English explanatory texts. Once the rupture with the UK was healed, the national atlas was published in English from 1982.

Kenya

We get a sharper sense of the postcolonial atlas by comparing the colonial with the postcolonial. The first *Atlas of Kenya* was published in 1959 when the country was a British colony. The "colonial gaze" is apparent in the description of landholdings categorized as "native reserves," "native settlement areas" and "native leasehold areas." The British Crown had alienated large amounts of land, and white settlers had established substantial landholdings. There is also a very detailed ethnic breakdown—vital information for a colonial power that maintained control through indirect rule. This tribal categorization made solid that which was more liminal with the formal depiction of tribal categories and their reinforcement in legal status. The British did not so much employ tribal identities but either invented them or reinforced them to maintain control through local elites described and empowered as "tribal leaders." On the more detailed maps, urban markers of colonial control, such as of private clubs and of prisons, were highlighted.

Kenya achieved independence in 1963 and became a republic in 1964. Atlases produced in 1970 and 2003 are now entitled *National Atlas of Kenya*, and while they still show similar material to the colonial era, especially scientific information such as rainfall figures, there is a subtle and sometimes not so subtle reframing. The historical geography of the region is now extended back beyond the colonial, and the colonial is even reinterpreted. In the postcolonial atlases, the Colonial Period is presented as just one era in a much longer, pre-European African history, and the European exploration is shown to be based on African labor:

> The production of the *National Atlas of Kenya* provides a fitting occasion to pay tribute to the achievements of the small body of enterprising Arabs and Europeans who first presented the geography of the country

to the world, and to the courage and endurance of the thousands of
African porters who made their journeys possible.[18]

Land categories are also changed to

> government land,
> trust land (formerly Native Reserves) and
> private land.

The postcolonial atlases are also framed as more than just economic inventories and ethnic affiliations. There is now more emphasis on the socioeconomic conditions of the population. The "natives" of the old colonial atlas are
now the citizens of the new postcolonial atlas.

Ghana

In the 1928 *Atlas of Ghana* produced while Ghana was a British colony, the
focus is on ethnic divisions within the country, its "Africannes" with images
of elephants and a highlighting of its role as export commodity producer. The
atlas portrays the country as part of an exotic Africa, as a source of primary
products with careful attention paid to racial and ethnic division, an essential knowledge to maintain control. A later atlas published in 1945 contains
maps of the British Empire situating Ghana in a wider world of British territories. Ghana attained independence in 1957. The first national atlas of the
new nation appeared in 1970. The title, *National Atlas of Ghana*, highlights the
project of nation-building. It depicts people not only by their tribal identities
but as socioeconomic units categorized by age and gender and depicted as
economic agents in their own right, because the economic geography now
extends beyond the narrow range of export crops in its depiction of local
markets and activities. Tribal division all but disappears in the postcolonial
atlases.

Vietnam

The atlases of Vietnam tell the story of colonial incorporation and postcolonial independence. While under French colonial rule, Vietnam was considered part of French Indochina, which also included Cambodia and Laos.
Maps and mapping of the territory were a vital part of how France depicted
and controlled its colonial holdings. The Saigon Post Office, built between
1886 and 1891 and still standing in the renamed Ho Chi Minh City, contains two wall maps made by the French, one of telegraph lines in Southern

Figure 4.10 Wall map in Ho Chi Minh City, Vietnam. *Source:* Photo by John Rennie Short.

Vietnam and Cambodia and another a map of Saigon (Figure 4.10). Several atlases were produced, including the *Atlas géneral de l'indochine française* published in 1909 and an *Atlas de l'indochine* published in 1928. The maps in these atlases depict the country as an economic resource with maps of rainfall as well as of mineral resources. The atlases depict territory divided into European controlled and indigenous. The country is framed as a resource base with extensive maps of mineral deposits, including tungsten, mercury, zinc, iron ore, silver, phosphates, coal and graphite; and all linked by roads, rails and telegraph and held together by the hierarchy of colonial power. In the key to the map of Hanoi, the governor-general's residence is afforded pride of place, mentioned first and recorded on the number index as 1. In 1970, an atlas of Indochina was published by the CIA. The atlas is very small with just 14 maps, with careful depictions of airfields. The 1996 *Vietnam National Atlas* marks the first major stocktaking of the socialist and independent Vietnam. It was produced in collaboration with the Russian Academy of Science. The introduction states that the atlas provides a

> panopticon view of the territory [...] marks a new era of the country [...] messenger of friendship between the Vietnamese people and the international community [...]
>
> shows a country is now under construction being developed fair and civilized.[19]

The atlas shows a country under construction after over a century of colonial exploitation, the large-scale destruction of the American War and the damaging U.S. embargo. However, the new atlas is no simple socialist propaganda: it outlines problems caused by unification, too rapid population growth, the thin spread of capital investment and the seriously defective system of subsidization. The atlas was published in 1996 as economic growth was faltering, so the pessimism reflects an external reality. The atlas was published at an inflection point just before the Vietnamese economy initiated sustained economic growth, rising living standards and reductions in poverty. The economic reforms toward a greater marketization referred to as *doi moi* were just beginning to take effect as the atlas was published. Written at a time when the economy was first being privatized, it reveals a searingly honest assessment of the problems of reforming a centralized and socialized economy. Written in the last decade of the twentieth century, the atlas reveals the ideological convergence of the postcolonial and the socialist with a wider globalizing world.

A Transect across the Ruptures

I began this chapter with a discussion of the 1916 *Geograficzno-statystyczny atlas Polski* (*Geographical and Statistical Atlas of Poland*) a cartographic certificate of independence while the territory was under the control of three separate empires, Austro-Hungarian, German and Russian. The subsequent twentieth-century history of Poland, from the Polish Republic to a People's Republic back to the Republic of Poland with changing territorial expressions, is embodied in successive editions of the national atlas.

The 1916 atlas prefigured the Polish state. The 1947 atlas, *Studium Planu Krajowego* (*Studies for the National Plan*) was produced at a pivotal time. It was printed in the same year as national elections were held. The Communist Party effectively came to power with vote-rigging and the outright persecution of the main opposition, the Polish People Party. The Communist-dominated government then exerted its total control, and Poland was incorporated into the Eastern bloc and the Warsaw Pact. The 1947 election was the closest Poland got to free elections until after the overthrow of the regime in 1989. The atlas of 1947 was published in Polish, English, French and Russian and probably completed before the total grip of the Communist Party. It stands then as a liminal text produced after the end of World War II in the immediate postwar period but before the icy grip of a repressive communist regime. The atlas was produced to supplement the National Plan drawn up in 1946 to rebuild the country after the devastation and chaos of World War II, the loss of lands in the east to the USSR and to incorporate the gains of territory in the west from a defeated Germany. The atlas is firmly rooted in the wider postwar European planning programs to rebuild shattered economies and badly damaged cities.

The atlas has wonderfully impressive maps that owe something to the inter-war expressionist style. It is as much graphical as cartographic. This graphic quality continues in more muted form in the 1973–78 *Narodowy atlas Polski* (*National Atlas of Poland*). The continuation of styles is indicative of an aesthetic legacy in the graphic arts. But the fact that this atlas was published only in Polish, though with maps keys and some text in English, is also indicative of the rupture of Poland from Western Europe during its period under Soviet domination. By 1993, after the country had escaped the grip of Communism and the Soviet Union had fallen, a new atlas was commissioned to represent the new state, *Atlas Rzeczypospolitej Polskiej* (*Atlas of the Republic of Poland*). It was published in English and Polish as a series of loose-leafed sheets in large boxes. As the surveyor general noted,

> The present atlas closes the epoch of censorship, ubiquitous propaganda and inaccessibility of much fundamental data. It is at the same time, the starting point for a new era and it will constitute a reference point for the future achievements of the new democratic Poland.[20]

The trajectory away from the Eastern bloc is reinforced in the text with numerous references to integration with Europe and the many maps that depict Poland firmly in the center of an extended Europe. Poland is carto-graphically shown as a central part of Europe, no longer the western edge of the Eastern Europe. The atlas is a celebration of the new nation and its distinctly European geopolitical location.

Chapter 5

NATIONAL ATLAS, GLOBAL DISCOURSES

The national atlas, as the name implies, is all about the nation-state: it represents the national territory and the national peoples, and it records, celebrates and sometimes idealizes the nation-state. Nationalist rhetoric drenches the nation-state. Yet this most nationalist of endeavors is also expressed in and through global discourses, making the most distinctly nationalist texts also one of the more universal. The global-national interaction in the national atlas is at the heart of this chapter. I will explore how the national atlas is a nexus between the national and the global. I will focus on three global discourses: international framing, language and cartographic conventions.

The Global Framing of the National

In the modern national atlas, usually one of the first maps depicted is a world map. This opening image invariably positions the nation-state in a global setting. The nation-state is depicted as unique but also as part of a wider world. The most distinctive global positioning was in the *Bol'shoĭ sovetskiĭ atlas mira* (*Great Soviet World Atlas*) where the world is represented on a map projection shaped as a five-star outline state with the USSR highlighted in bright red (Figure 5.1). The whole world is recast in the form of the Soviet emblem with the USSR in prime location, suggestive of a world about to be dominated by the workings of history and forces of world communism into a USSR-shaped world.

In most cases, the initial world map is drawn to highlight the nation-state. Numerous atlases frame the world map from the position of their nation-state. In some cases, this involves novel perspectives. In the 1954 edition of the *Atlas de la República Argentina*, for example, the world is represented in the Mercator projection. However, it is centered on longitude 60 degree west, not the usual centering on 0 degree in Greenwich, England, which would have offset Argentina from the center of the map. So, a world map is used that frames

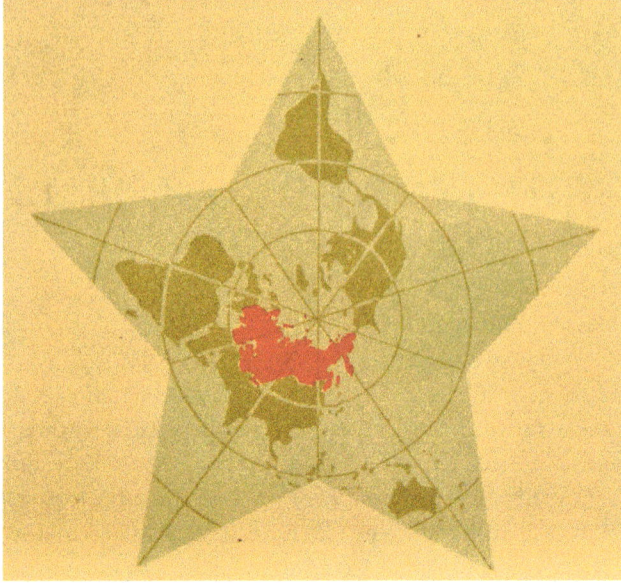

Figure 5.1 Situating the USSR, *Bolshoi sovietski atlas mira* (*Great Soviet World Atlas*), 1937. *Source:* Photo by John Rennie Short.

Argentina at the center of the world with great subtlety. In the 1974 *National Atlas of Canada* several maps use polar projections, centered close to the North Pole rather than Mercator projection. Canada is at the center of the world from this perspective. The world map in the early pages of the *National Atlas of Jamaica*, published in 1971, uses a world map projection with Jamaica at its center. The newly independent country is recentering the world after being on the colonial edges of the British Empire. In the Romanian national atlas of 1974, *Atlas Republica Socialistă România*, the first map is a world map centered on Romania set off from the surrounding dull, yellow-colored land masses by bright red coloration. The 1979 *Atlas de Venezuela* used a double hemispheric map of the world, which looks modern and scientific but also has the advantage of framing the country in the center of the map. The 1998 *National Atlas of Sri Lanka* has a world map on an Azimuthal Equidistant Projection. The title says it all: "World Centered on Colombo." Across the title page of the 1999 *Atlas of Saudi Arabia* is the country's outline in light green against a dark green globe.

The centering is also apparent in the maps of global flows. Many world maps depict the flows to and from the nation-state. The flows depicted in this global centering also reveal the key international trading relations of the nation-state. The 1990 *National Atlas of Japan* contains maps of Japan's economic geography. Several world maps, all centered on Japan, show the import of iron ore by country. Australia is revealed as a major supplier of this vital ingredient to Japan's rapid economic growth. The type of flows used is revealing. In the 1993 *National Atlas of UAE*, maps of linkages include the origin of migrants to the United Arab Emirates (UAE), a region heavily dependent on foreign labor, both skilled and unskilled. The 1999 *Atlas of Saudi Arabia* has two world maps of Saudi connections with the wider world. One depicts the exports of crude oil by country; the main flows at the time are to China and the United States. The other depicts pilgrims arriving in 1994 by continent and country. The 2000 *Atlas Nacional do Brasil*, for example, depicts the global flows of capital into the country.

The very publication of the national atlas can also signal a geoeconomic and geopolitical repositioning in the world. A plan to publish a national atlas of China was part of the 12-year plan drawn up in 1956. Four volumes were planned. However, the work faltered during the turmoil and disruption of the Cultural Revolution (1966–76). In 1980, the Chinese government authorized the atlas to be completed. Its publication, only two years after Deng Xiaoping announced an open-door policy to foreign investment, was one of the many ways that China signaled its openness to the wider world. An important goal of the national atlas was to provide information to the outside world. The national atlases of China produced after 1994, for example, all aim to depict advances in science and technology, economic growth and the building of socialism with Chinese characters. The *National Economic Atlas of China* was published in English in 1994 in Hong Kong. The very first map features a bright, glowing, yellow-colored China that, by the way, includes Taiwan, Hong Kong and the expansive claim to the South China Sea. China is now depicted as part of and at the center of the international arena after years of self-imposed exile (see Figure 5.2).

There is an epistemic framing and reframing, a positioning to wider political and intellectual frameworks. It is perhaps easiest to express this through a before-and-after repositioning. Produced in 1966 under the military dictatorship that ruled Brazil from 1964 to 1985, the *Atlas Nacional do Brasil* has a traditional suite of maps that concentrates on economic performance with maps of imports and exports and on population movement into the Amazonia interior. This national atlas is devoted to modernization and expansion and expressed in maps celebrating a dynamic capitalism of economic growth and of the projection of state power into the interior. The 2000 edition of *Atlas Nacional do Brasil* was produced after the fall of the military dictatorship. Introductory essays by radical scholars such as Milton Santos and Bertha Becker expand on themes of global geopolitical restructurings. The maps are informed by a more critical political economy approach to state power and capitalism compared to the apologist tenor of the previous atlas. The maps

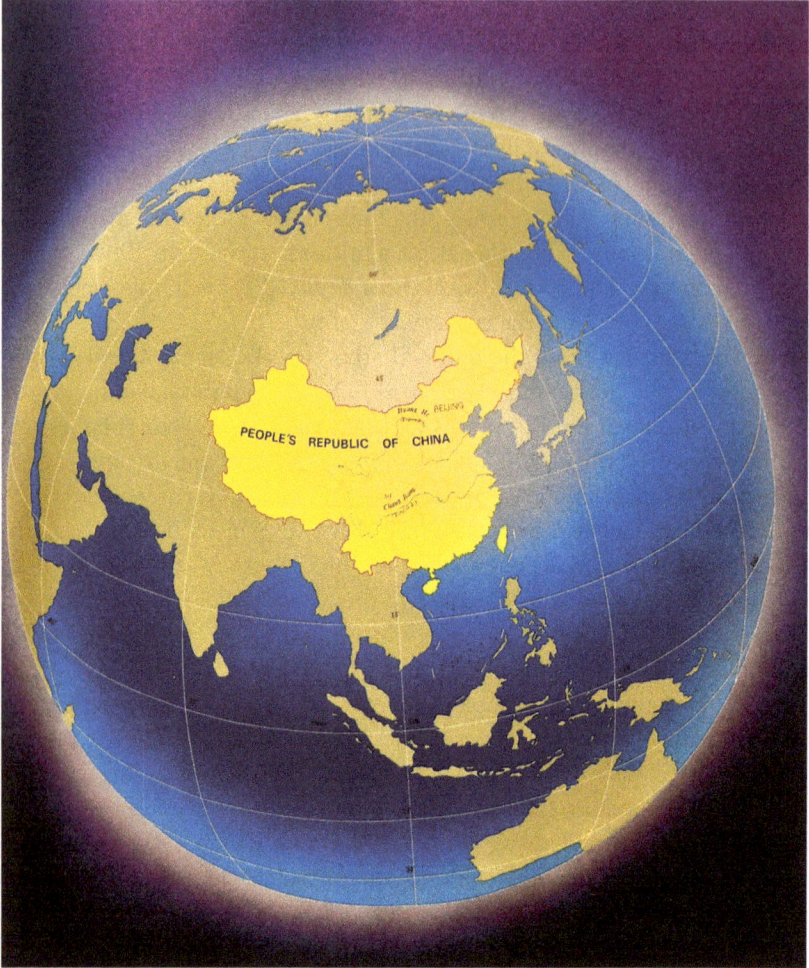

Figure 5.2 A Chinese-centered world, *National Economic Atlas of China*, 1994. *Source:* Photo by John Rennie Short.

of investment flows recall the capitalist investment flow of the *Bol'shoĭ sovetskiĭ atlas mira* (*Great Soviet World Atlas*). See Figure 5.3 compared with Figure 4.3. A more critical interpretation of economic growth and modernization is revealed by maps that deal with infant mortality and public health, topics absent from the atlases of the junta days. While the previous atlas depicted the movement of people into Amazonia as a form of development, the more critical approach of the 2000 atlas depicts maps of assassination of rural workers and of peasant conflicts. The space of Brazil is reframed and reinterpreted by a more critical political economy approach.

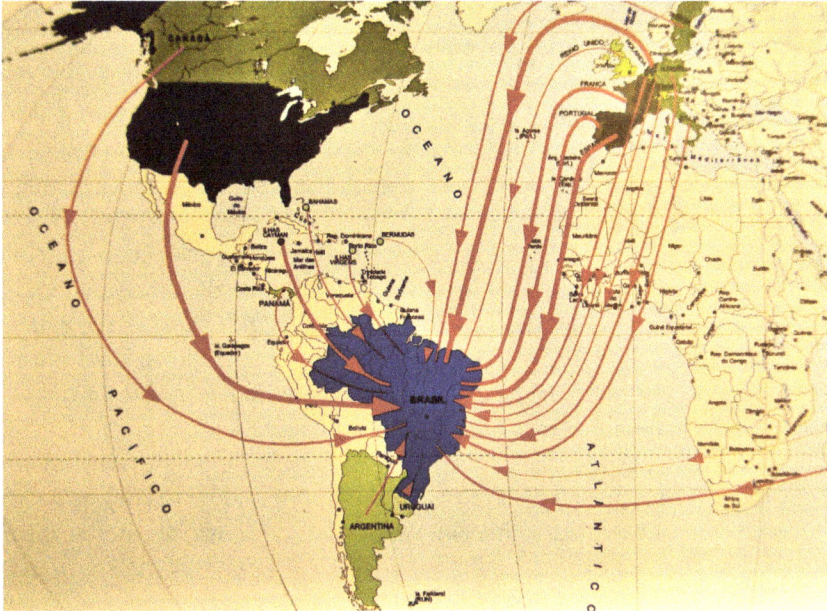

Figure 5.3 Capital flows into Brazil, *Atlas Nacional do Brazil*, 2000. *Source:* Photo by John Rennie Short.

Sometimes the repositioning is brief. The *Atlas of Canada*, first published in 1906, focused on Canada as a new nation. However, the first map of the second edition, published in 1916 during World War I, was a world map with Canada colored red as was the UK. At a time of patriotic fervor, the atlas reconnected with its imperial linkages. But by the time of the 1958 *Atlas of Canada*, a world map depicted Canada as part of the Commonwealth, the Colombo Plan, NATO and the United Nations. Canada was reframed toward a wider world rather than just the British sphere of influence.

Maps in the national atlases situate the nation-state at different regional and world scales. The 1967 *Atlas de Colombia*, for example, highlights Colombia in a series of maps. Colombia in the world, Colombia in South America and then two further maps overlay the country's outline over Central America and the United States. In the second of these maps, Colombia is outlined in green right in the middle of the continental United States, an interesting choice of overlay suggestive of the close ties between the two countries during the height of the Cold War as the United States sought to counter leftist movements on the continent, and Colombia elites sought support from the United States against populist movements. In the 1993 *National Atlas of the United Arab Emirates*, the emirates are depicted from space and then in a series of maps of the Islamic world, Arab world and the Gulf.

Sometimes the global positioning is revealing. In contrast to the geographic overlay of Colombia and the United States suggestive of close ties—one country literally on top of the other—the 1978 *Atlas de Cuba*, produced after Castro came to power, shows global flows of imports and exports to and from Cuba. The main trade links are with Europe and especially the USSR and Eastern Europe. There are very few links with South and Central America, a small amount of trade with Canada and no links with the United States. It is a Cuba primarily connected to Europe and the communist world and cut off from the United States by formal embargoes. Compare this with the prerevolutionary links depicted in the 1949 *Atlas of Cuba*, where the island is depicted in figurative form sitting in the sea with export arrows pointing out, their size indicating the size of the trade. The biggest link, by far, is with the "Estados Unidos." The different national atlases highlight the changing international frame of Cuba from an economic colony of the United States to a socialist nation with many links to Europe and with the then communist nations of the world.

A global repositioning is evident in comparing precolonial and postcolonial atlases. The *Atlas des Cercles de L'A.O.F* was a series of atlases of the French territories in West Africa. AOF stands for "Afrique occidentale française." One volume, *Fascicule VII*, was devoted to Senegal. The atlas, published in 1925, positions Senegal as part of a French territory with no linkage to the world outside of France and its territories in West Africa. Maps show administrative units, the racial-ethnic mix and the commercial links of telegraphs, mail links and roads. In the 1977 national atlas produced after independence in 1960, *Atlas National du Sénégal*, the maps at the beginning of the volume position Senegal in a much wider world. The global map "Le Sénégal dans le Monde" shows diplomatic connections as well as airline and ship networks that link Senegal to a wider world. And another map, "Le *Sénégal* dans les relations Interafricaines," shows similar connections with countries across the African continent. The atlas inserts Senegal into wider African and global contexts, repositioning the territory from its cartographic presentation as an element in the French Empire to an independent country with its own connections across the continent of Africa and the wider world. Just as much as the series of maps of Senegalese resistance to the colonial conquest, these global maps highlight the distancing from France and the rupture with the colonial past.

All the modern national atlases depict the nation-state not only as a distinct territorial unit but also as part of a wider world. The international framing of the nation-state is expressed in how the state is cartographically presented, the type of connections that are recorded and mapped, and the epistemic framework adopted. Despites it primary national concern, the national atlas employs cartographic devices and draws on often changing intellectual frames that link the nation-state to wider global discourses.

The Language of the Atlas: Text

National atlases have a significant amount of text. Words are used to introduce the project, to frame and inform the maps and to guide the reader through the images. A significant feature is the growing dominance of the English language as the twentieth century progressed. The most national of texts was often expressed and encased in one global language—English.

Consider the case of Finland. The 1899 *Atlas öfver Finland* can be considered one of the earliest examples of the modern national atlas: it was published in the two languages of the country. Swedish was the language of the elites and the main language used in the higher education system while Finnish was the vernacular language of the majority of Finns. No English appears in the text or in the keys. The main title of the atlas is Swedish, a reminder of the continuing legacy of Swedish linguistic dominance for Finnish intellectuals. By 1910, the second edition of the *Atlas öfver Finland* was published in three languages, Finnish, Swedish and French. The use of French internationalized the atlas to a wider audience. In 1922, an independent Finland endorsed a bilingualism with Finnish and Swedish as the official languages of the nation-state. At the time, around 13 percent of the population spoke Swedish and the language still had a strong cultural legacy in the academic communities. But in the third edition of 1925, the preface is in Finnish, English and Swedish. French is dropped and Swedish is used but downplayed and no longer the lead language in the title. By the time of the 1960 edition, Finnish was prioritized in larger print size while many of the keys to the maps were only in Finnish and English. In the remainder, the order is Finnish, English and Swedish. Today, Finnish is the first language of 87 percent of Finns, while Swedish is the first language of only 5 percent and the second language of 44 percent. Finland is officially a bilingual country with Swedish-speaking Finnish citizens having the right to communicate to state authorities in their own language. However, the language of the national atlas reflects the linguistic practices of most Finns and the dominant form of communication of the global epistemic community, English.

The changing linguistic usage of the successive atlases of Finland reveals the shift toward the use of "national" languages, in this case Finnish as well the emergence of official policies of bilingualism, in this case Finnish and Swedish, and the growing usage of English. The changing linguistic frame for the subsequent editions of the national atlas of Finland highlights a feature common to many other national atlases, the integration of English into the text.

Over the course of the twentieth century, English became the dominant language of global epistemic communities. The widespread diffusion of

English was promoted by the British Empire and the post–World War II economic and cultural power of the United States. English was the dominant language of both the nineteenth- and twentieth-century superpowers and was reinforced by more recent economic and cultural globalizations. English was not just a global language as Spanish, but had become *the* global language, the lingua franca of global interactions. The use of English was utilized as part of globalization strategies in many non-English-speaking countries. Its usage was especially pronounced in the epistemic communities where most scientific, technological and academic information is expressed in English. Over 80 percent of all information stored in electronic retrieval systems is in English. The use of English as a written global form of communication is particularly visible in communities of knowledge. English has become the language of global intellectual discourse and the dominant language of intellectual communities (*episteme*, in Foucault's notion) involved in the production, reproduction and circulation of knowledge. Publishing works in English has become a scientific habit of scientists around the world. In academic texts, such as a national atlas, there is a trend toward linguistic homogenization.[1]

In some cases, the use of English was part of the colonial legacy. Former British colonies such as Ghana, Jamaica, Kenya, Pakistan, Sierra Leone, South Africa and Sri Lanka all used English as the main and often the only language in their national atlas. It was not just the case of continuing colonial legacy. Many of the postcolonial states such as Ghana, Kenya, Pakistan, Sierra Leone and Sri Lanka had multiple language communities. In the case of Sri Lanka, three different languages were employed, English, Sinhalese and Tamil. English could be used as a language unifier among the different language communities where language usage also aligned with divisive race, ethnic and religious differences. In the language map of the 1999 *Atlas of Pakistan*, 12 different major language groups are noted, including Balti, Baluchi, Kashmiri, Punjabi and Pushto. The accompanying text to the map notes, "All educated Pakistanis are trilingual having studied Urdu and English besides their regional language" (see Figure 5.4). For the postcolonial multilingual societies, English acted as a convenient common language form, especially for the more affluent and better-educated elites.

In some instances, there was linguistic resistance. The first national atlas of Israel, the 1956 *Atlas Yisra'el*, was published in Hebrew. Later editions of the *Atlas of Israel* in 1970 and 1985 used both English and Hebrew in the text and the 2011 *New Atlas of Israel* was published in both an English and a Hebrew edition. The 1959 edition of the *National Atlas of India* was in Hindi, including the key to maps and notes, although there were 25,000 words of English explanatory text. This early postindependent edition of the national atlas was

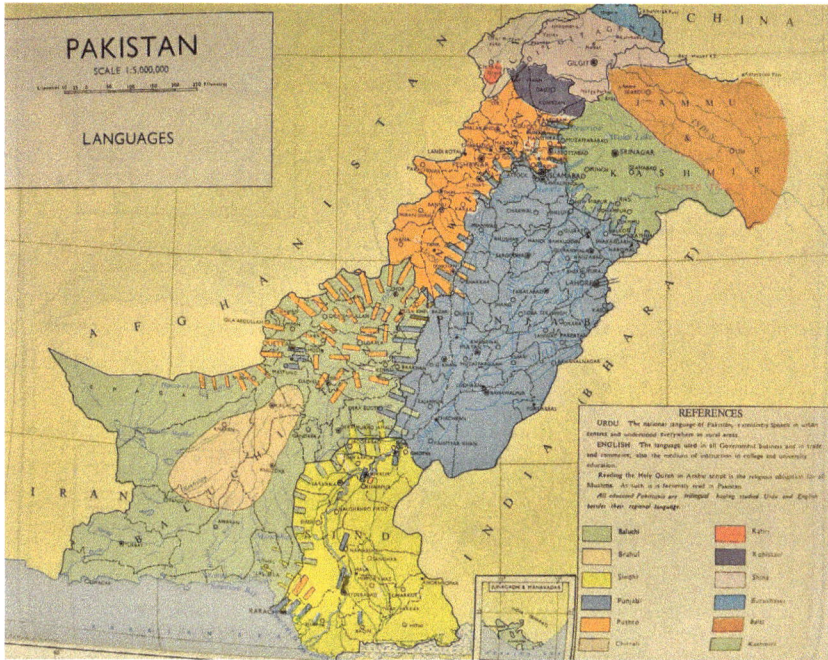

Figure 5.4 Languages of Pakistan, *Atlas of Pakistan*, 1997. *Source:* Photo by John Rennie Short.

self-conscious distancing from the language and measurement system of the former imperial power. Not only was Hindi used but the metric system was also adopted. By the time of the massive 1982 edition of the *National Atlas of India*, which includes 330 plates in eight volumes, all the text was in English. The use of English was not a return to neocolonial ways but a recognition that the atlas had to be published in the international language of English to have wider international relevance.

Some nation-states resisted the growing use of English. The national atlases of South and Central American countries such as Argentina, Brazil, Colombia, Cuba and Venezuela all resisted the use of English. This was, in part, because of their use of the already globalized languages of Portuguese and Spanish. With less formal British cultural dominance to act as legacy and strong resistance to the linguistic imperialism of the giant neighbor to the North, the national atlases of South and Central America maintained their linguistic independence from the growing use of English.

Russian was the only language used in the 1937 *Bol'shoĭ sovetskiĭ atlas mira* (*Great Soviet World Atlas*). The Russian resistance to English usage continued long after the fall of Communism. The 2004 *Nats̆ional'nyĭ atlas Rossii v chetyrekh tomakh* (*National Atlas of Russia in Four Volumes*) was printed only in Russian. The Eastern bloc countries tended to resist the sweeping use of English. The *National Atlas of Poland* produced between 1973 and 1978, when the country was part of the Warsaw Pact, was published only in Polish. In contrast, a Romanian government seeking foreign loans was more amenable to international languages use. The 1974 *National Atlas of Romania* was published in Romanian, English, French and Russian.

English was used in the atlases in the Middle East. The 1973 *Atlas of Iran White Revolution*, produced while the shah was still in power, was printed entirely in English. Arabic was the only language used in the 1975 *Atlas of Saudi Arabia* but by the time of the 1999 *Atlas of the Country of Saudi Arabia*, the text was still dominated by Arabic but the titles and keys to the maps were in both Arabic and English. The 2006 *al-Aṭlas al-waṭanī al-Qaṭarī* (*Qatar National Atlas*) was published in both Arabic and English, although the text was organized from right to left, Arabic style rather than the Romanized left to right.

Nowhere was the use of English usage to connect with a more international audience more apparent than in the case of China where the return to the global epistemic community was signaled not only by the publication of national atlases but by their publication in English. The *Population Atlas of China* (1987), *The National Economic Atlas of China* (1994) and *The National Physical Atlas of China* (1999) were all printed in English. As the 1994 volume noted, it was not only a document to plan economic development but also a way for the rest of the world to understand China. And the rest of the world was increasingly using English as the language of its understanding. Even Vietnam, a country shattered by the American War, used both English and Vietnamese for the 1996 *Vietnam National Atlas*. The need to connect with a wider world and attract foreign investment necessitated the use of the global language. The growing dominance of English meant that the most national of texts were often framed in the global language of English. The National *Atlas of Japan* was published in both English and Japanese. The reasons cited in the text for this publication included the need to inform government policy, to provide data for researchers and private business, to aid in education and to provide foreigners with knowledge of the country and its people. A revised edition was published in 1990, and in the preface the prime minister noted that the volume would provide a more global understanding of Japan and the Japanese. The publication in English was vital to achieve the aim of this national atlas and the national atlases of other emerging countries. Because the national atlas was a showcase of the country to the wider world, it was necessary to make it legible for a global audience.

The Language of the Atlas: Maps

National atlases represent the territory of the nation-state. They do so in two types of maps. There are the general topographic maps of territory. These include maps of the nation and constituent regions depicting numerous features, including altitude, hydrology, transport links, towns and boundaries. These maps use a common language. Altitude, for example, is generally presented by several means including contours—lines of equal altitude—shading and less frequently, "hachuring," which is the use of small lines to indicate steepness. Towns and cities are represented by dots and circles with their size a measure of the population. By convention, water areas are colored blue. In other words, there is standardization of the cartographic language across the atlases of different nation-states. Maps can be easily read by those versed in cartography.

Figure 5.5 is map from the 2004 *Nats ional'nyĭ atlas Rossii v chetyrekh tomakh* (*National Atlas of Russia in Four Volumes*). This atlas was published only in Russian and yet we can "read" the map even if we do not understand Russian. Altitude is expressed by contour lines, shading and spot heights so that we can literally "see" the conical hill beside the coast. We can make out rivers, roads, buildings and even an airport. We can even discern the depth of the water. The map is legible.

Topographic maps in atlases, whether of the entire nation-state or individual regions, are part of a global metric. All the general maps in the national atlases use the global grid of latitude and longitude. This grid turns the singularity of different and unique places into the smooth global surface of mathematical space. Longitude is measured 180 degrees east and west of the prime meridian in Greenwich, England. Latitude is measured 90 degrees north and south of the equator. Each degree of latitude and longitude is further divided into minutes and seconds. Today, reliable global positioning systems (GPSs) can provide pinpoint accuracy. I grew up in the village of Tullibody in Scotland. The village has a long, centuries-old history, and its name derives from the Gaelic. But this rich historical geography is turned into the austere form of the coordinates 56.1334° N, 3.8377° W.[2] All the places in the world submit to similar coordination. Look again at Figure 5.5 and you can see how a grid of latitude and longitude encases the national territorial depiction of Russia in a global grid.

The topographic maps of the national atlases are integrated within a global measurement system. This was not a given, but the result of scientific measurements and scientific conventions. The search for the accuracy of the grid was a long struggle. In 1669–70, the Frenchman Jean Picard was able to accurately measure the size of the earth by extrapolating from the distance of

Figure 5.5 Topographic map *Natsional'nyĭ atlas Rossii v chetyrekh tomakh* (*National Atlas of Russia in Four Volumes*), 2004. *Source:* Photo by John Rennie Short.

one degree of latitude measured in the French countryside. However, similar surveys of a degree of latitude undertaken in Peru and Scandinavia had different values. The earth is not a perfect sphere but flattened at the poles. A degree of latitude is shorter closer to the equator than the poles. The earth is more of an oblate than a sphere. In 1768, the English surveyor Charles Mason, one half of the Mason-Dixon partnership, compared his measurements with those done in Scandinavia and concluded that the contrasting calculations were the result of the difference in the earth's density. A series of careful measurements were made by British surveyors of a hill in Scotland, Schiehallion. In 1775, Nevil Maskelyne read the results to the Royal Society to confirm the effect of mountain mass. The measurements ranged from the hills of Scotland and the mountains of Peru to the plains of India and the Maryland tidewater. Scientists were thus able to work out the differential density of the earth's core through surface mapping, a genuinely impressive achievement before deep-core probes or sonic measurements. It is an interesting tale that reveals much about how we came to a better understanding of a less-than-perfect sphere in all its misshapen complexity.[3]

But not just scientific measurements were needed. While the equator is the obvious place to base the measurement of latitude, longitude is more subjective. There is no "natural" prime meridian on which to base zero-degree longitude. In the eighteenth and nineteenth century, many countries used their capital city as their prime meridian. Up until the late nineteenth century, U.S. maps, for example, used first Philadelphia, then Washington, DC, as their prime meridian. The name of Meridian Hill Park in Washington, DC, remains as a legacy. A universally agreed prime meridian was only conceded in 1884. The decision was spurred by the need to have an agreed-upon time for the train system to work effectively and to achieve an international standard.

This standardization of longitude means that all places are now located on an agreed-upon grid. Most topographic and general maps in national atlases are framed with degrees of latitude and longitude in the border. The unique territories are depicted in a universal metric. The grid of latitude and longitude was made possible not only through improvements in measuring the earth but also by the emergent global community of science.

The modern national atlas contained topographic maps but also thematic maps. There are maps that depict specific themes such as population density, sites of industrial production, language groups, etc. Again, there is a standard repertoire of techniques. The 1899 *Atlas öfver Finland* set the standard for subsequent atlases. It was presented to the Berlin Congress of the International Geographic Union (IGU) in 1899 and became a template for later atlases and guided the IGU Commission on National Atlases, established in 1956, that established the following requirements for the national atlas:

- comprehensive content aiming at completeness,
- series of uniform, comparable and complementary maps of high scientific quality,
- practice oriented,
- useful for the general public,
- five major topics (nature, population, economy, culture and politics),
- a manageable size (max. 40–50 cm × 60–70 cm),
- uniform scales to enable comparison,
- explanatory texts to enhance understanding of the maps possibly with graphics.

We can see all these items as the ideal case, the aspiration if not always the result. There was some tension in the pursuit of the goals. The national atlas did aim at comprehensiveness that meant they often ballooned in size beyond a "manageable size." The size of the project and indeed the size of the resultant text made them less "useful for the general public" and more useful only for storage on the outsized shelf of the reference section. However, these elements generally *do* mark the national atlases of the twentieth century: large, comprehensive volumes with similar suites of maps, slotted into the global grid and a similar range of thematic issues such as physical environment population economy, culture and less frequently politics.

The cartographic language of thematic maps in the national atlas both reflects and embodies a global discourse of scientific practices. Figure 5.6 from the 1988 *National Atlas of Ethiopia*, for example, depicts the climate classification of Ethiopia. Similar types of diagrams and maps can be found in many national atlases. Figure 5.7, from the 1962 *Atlas of Kenya*, shows rainfall and humidity. The similarity of the type of data collected, classified and displayed is indicative of shared global scientific practices that make the maps of the national atlas legible to a more global audience. In the next chapter, we will come across geology map from the 2009 *National Atlas of Korea* that draws upon standardized geological classifications (see Figure 6.8). Weather, climate and geology as well as other physical phenomena are expressed in national atlases in the shared classification system of a global science.

It is not only the physical environment that is represented in a shared cartographic language. Population data are often depicted with shading in what are defined as choropleth maps. Figure 5.8, from the 2003 *National Atlas of India*, depicts the Muslim population of India. Notice how easy it is to identify the areas of concentration of Muslims. Maps that show variations in, for example, education levels or health status across regions use devices such as circles or vertical bars on the map to represent the different values. Also common is the representation of flows across space with width reflecting size of flow. Figure 5.9, from the 1979 edition of the *Atlas de Venezuela*, depicts flows of passenger traffic from provincial centers to the capital city of Caracas. The structure of the flows is suggestive of a country dominated by the urban primacy of the capital city.

There is a standard repertoire of thematic mapping techniques used in the different national atlases. A series of regional maps, for example, can highlight weather patterns at different seasons such as shown in Figures 5.6 and 5.7. So, while the national atlas represents a unique territory, the technical means used to represent spatial variations have a similarity that amounts to a common cartographic language across different national atlases. This common language makes them legible and makes possible an understanding across different national audiences. Just a few words in English in the key to most thematic maps make them more universally understood to the global

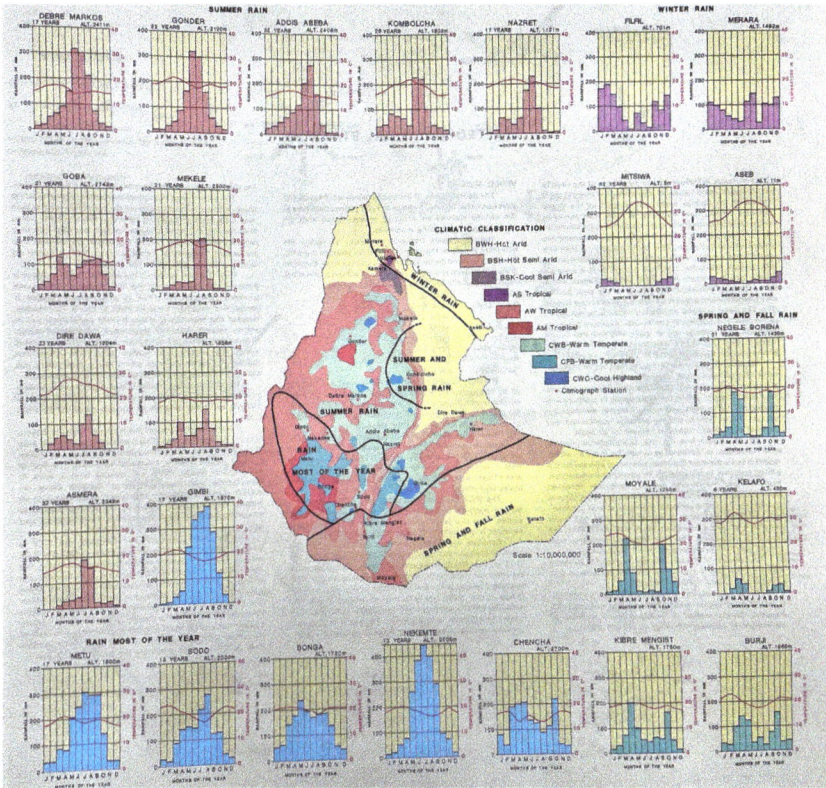

Figure 5.6 Climate classification, *National Atlas of Ethiopia*, 1988. *Source:* Photo by John Rennie Short.

English-reading audience. The national atlas is written in languages legible to a global audience.

The standardization of the maps, both topographic and thematic, in the national atlases results from the exchange of ideas as mapmakers learn from, copy and seek to improve upon the work of previous mapmakers in earlier atlases. There is a circulation of knowledge. Sometimes the international exchange was aided by direct collaboration in the form of aid. The USSR Academy of Science helped in the production of the national atlases of Cuba and Vietnam, Polish experts collaborated on the atlas of Afghanistan, a Canadian development agency provided support and assistance to the atlas of Ethiopia, Swedish experts on the atlas of Botswana and French scientists with CNRS (Centre national de la recherche scientifique) aided with the atlas of Senegal. The French influence persisted even after independence. The 1977 *Atlas National du Sénégal* has a significant neocolonial legacy. Four of

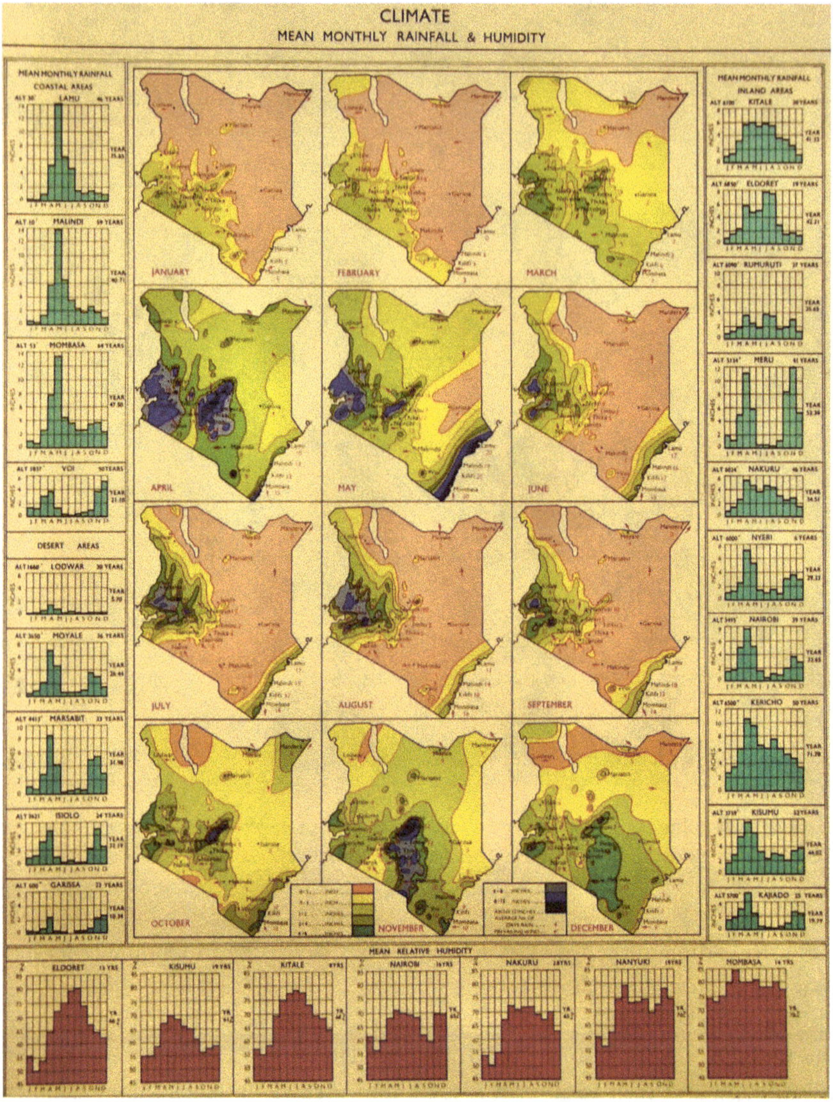

Figure 5.7 Rainfall and humidity, *Atlas of Kenya*, 1962. *Source:* Photo by John Rennie Short.

the senior committee members were based in French universities and of the 13 institutions collaborating in the making of the national atlas, 11 were in France or in CNRS. Jamaica's first national atlas was produced in 1971 under the auspices of a United Nations Special Fund. And German experts from the German Agency for Technical Cooperation worked on the 1983 atlas of Liberia. The 1993 atlas of the UAE hired academics, based in Kuwait

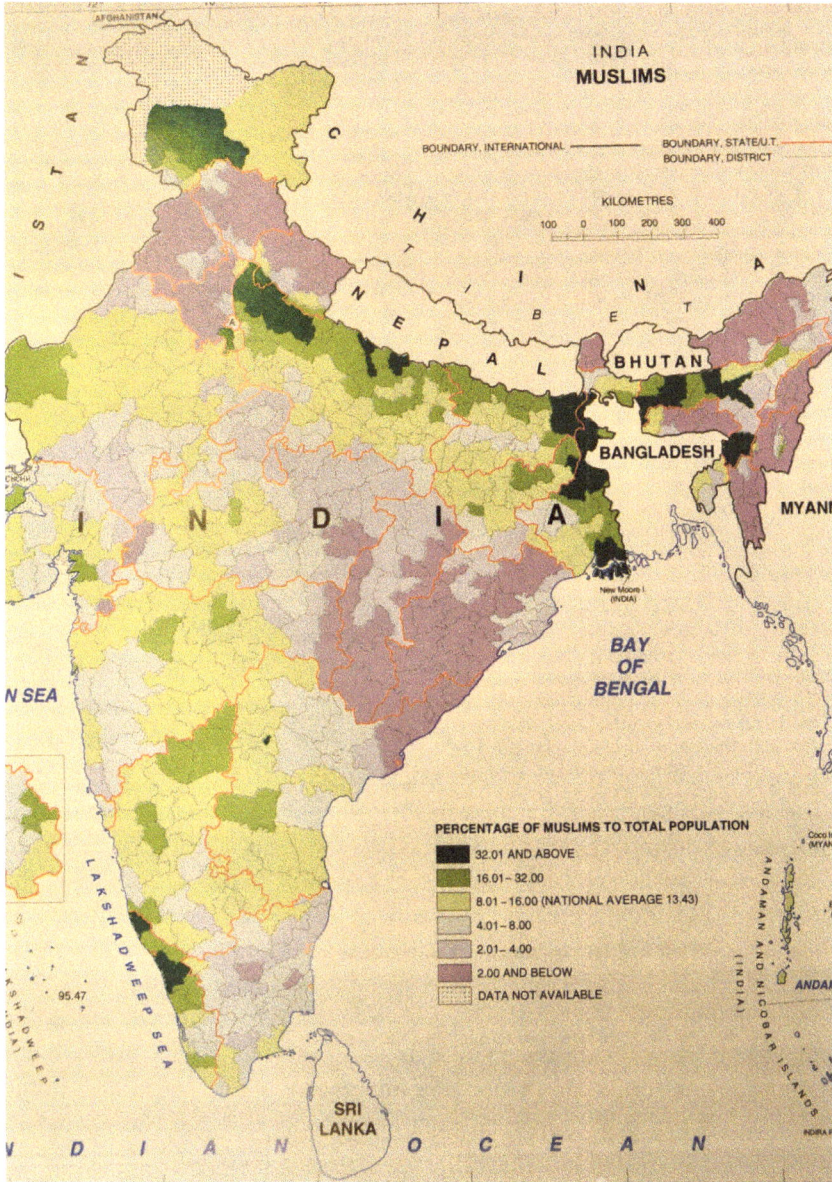

Figure 5.8 Map of Muslim population, *National Atlas of India*, 2003. *Source:* Photo by John Rennie Short.

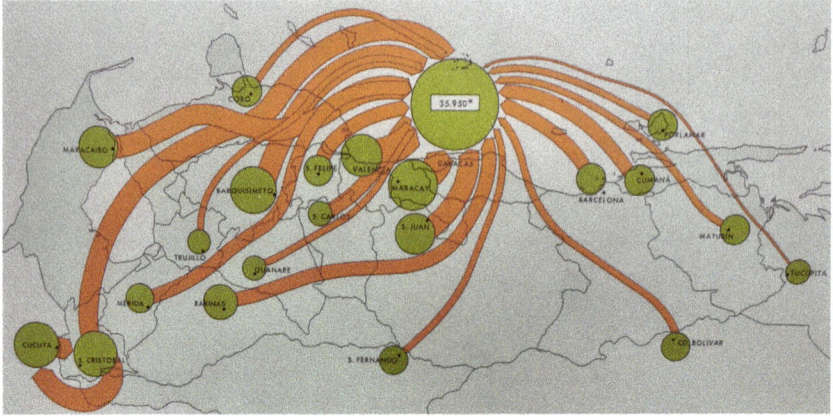

Figure 5.9 Passenger flow map, *Atlas de Venezuela*, 1979. *Source:* Photo by John Rennie Short.

and Lebanon, and employed an English consultancy company. There are elements of an epistemic neocolonialism, and some nations were so acutely aware of the relationship that they sought to distance themselves. The preface to the 1988 *National Atlas of Sri Lanka* felt it had to note that it was an entirely Sri Lankan product, and the preface to the 2003 *Atlas of Kenya* noted that this was a different edition because it was prepared by local staff using modern personal computers and workstations. There is an unequal exchange: African experts were not involved in the production of atlases in France or Germany or Poland. Because national atlases are the product of international collaboration both direct and indirect, the result is a standardization of the atlas form. The standardization is reinforced using international experts and consultancy who aid the circulation of similar atlas forms around the world.

The textual framing of the maps also exhibits a common pattern. The shift toward multimedia text, with maps, diagrams, figures and texts all on the one page, became a common feature as atlas designers learning from other atlases drew on similar and sometimes improved images. The images in the different national atlases take on such a similarity that the exceptions are noticeable. The atlas of Japan normally employs the more universal language of maps and textual design. However, in the *National Atlas of Japan* published in 1977 there are several flow diagrams of internal trade. The ones of rice and crude oil have a style very reminiscent of traditional Japanese maps in the 1,200-year-old Gyoki tradition that have a marked geometric form. These maps have a smooth sinuosity of a geometric form that recalls flowing water. The map of rice flows across Japan is a cartographic echo of this ancient tradition (see Figure 5.10). They are distinctive because they present one of

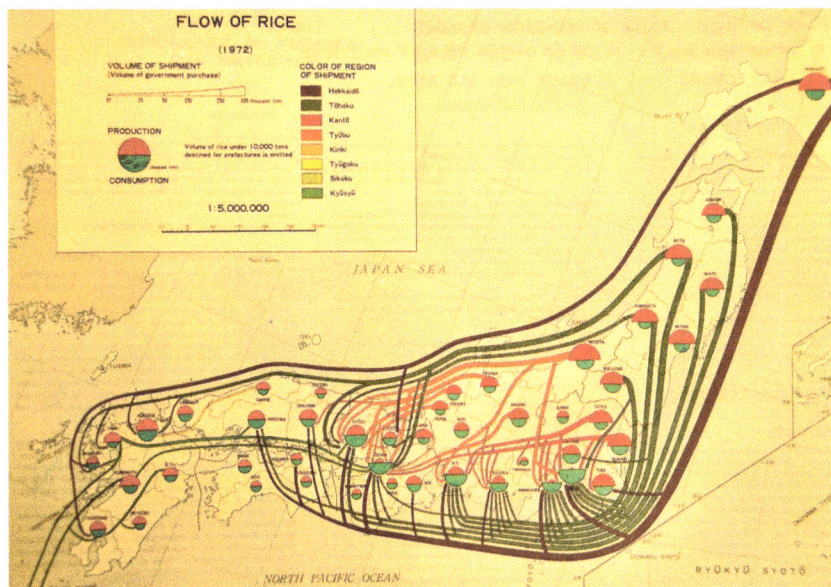

Figure 5.10 Flow of rice, *National Atlas of Japan*, 1977. *Source:* Photo by John Rennie Short.

the few examples of an older vernacular tradition of mapmaking among the global uniformity of most modern national atlases.

The national atlas, one of the most distinctly national of texts, was shaped by global discourses and informed by universal cartographic and linguistic conventions. The modern national atlas emerged as a distinctly national text in a shared global epistemic.

Chapter 6

THE PHYSICAL WORLD OF
THE NATIONAL ATLAS

The national atlas is a state-sponsored, scientifically informed territorial imagining. The state is a political arrangement that claims monopoly power over the land and the people of a specific territory. The national atlas is an embodiment of this claim to authority, a technical performance of power and legitimacy. In this chapter, I will look at the land. The next chapter I will consider the people.

The national atlas depicts the physical space of the nation-state. I will look at the constants as well as the changing notions of what constitutes this physical space. I will consider four main themes: the origins of the science of the national atlas; how the national atlas involves the coproduction of science and state; how the national atlas condenses national imaginaries; and the contested discourses of the national atlas, paying particular attention to boundaries, toponymic issues and the changing understandings of the physical nature of territory in the Anthropocene.

Origins

If we agree that the national atlases contain information, as well as imaginings, then what exactly is the type of information considered suitable. In the traditional national atlas, there is a usual sequence of maps. A map of the nation-state in the world is often followed by maps of the national territory. Then there are the thematic maps that depict specific topics. Some sense of the scope is given in the second edition of the *National Atlas of India* produced in 2003–9. It is one of the largest national atlases with 243 separate maps in 10 volumes. Table 6.1 lists the comprehensive list of topics. These are found in similar, though often smaller variants, in most national atlases. Where and when did this intellectual ambition arise?

Table 6.1 Volumes of *National Atlas of India* (2003–9).

Volume I	General and Political (20 maps)
Volume II	Physical and Geomorphological (24 maps)
Volume III	Climate and Water (26 maps)
Volume IV	Land and Agriculture (32 maps)
Volume V	Population (21 maps)
Volume VI	Social (28 maps)
Volume VII	Infrastructure, Economic, Industries (23 maps)
Volume VIII	Environment, Science and Technology (20 maps)
Volume IX	Health (19 maps)
Volume X	States and Union Territories (30 maps)

A long tradition of cosmography

The ambitious range of topics in the national atlas is the continuation of a long tradition. The national atlas of the twentieth century owes much to the cosmography of earlier centuries. Cosmography emerged in the sixteenth and seventeenth century in Europe as an attempt to encompass the world of knowledge with a knowledge of the world.[1] It has roots in the work of Persian and Arab scholars of centuries earlier. Cosmography had the goal of trying to understand the totality of the world, the macrocosm of the celestial order as well as the microcosm of terrestrial arrangements. Based part on measurement and empirical inquiry, it also had a mystical basis. In the European Renaissance, cosmographers included Sebastian Munster (1488–1552) and John Dee (1527–1608) who sought to write a geographical understanding of a known world that was growing in size and complexity as Europeans extended their knowledge of, and connections to, a wider world. These writers sought to impose order on an unruly world when the difference between what we now consider science and magic was not so stark nor as sharp as it is now. Religious fervor, mysticism and downright wackiness enter the scene alongside careful description. Consider the case of Sebastian Munster. Born near Mainz in 1505, he went to Heidelberg to enter a Franciscan order. Two years later he went to Louvain to study mathematics, geography and astronomy. In 1509, the order sent him to the monastery of St. Katharina in Rufach in the Vosges Mountains to study geography, mathematics, cosmology and Hebrew. Hebrew scholarship and geographical understanding were recurring interests in Munster's professional life and all part of understanding God. For him, there was no paradox between established religious faith and the new forms of knowledge. Munster was a prolific writer. He published 80 books on theological subjects and translated a Hebrew text of the Bible into Latin. Munster continued his travels, extending his knowledge by traveling to centers of intellectual activity. He stayed in Tubingen and in 1518 moved to Basel. In

1524, he was appointed professor of Hebrew at Heidelberg University, where he also lectured on mathematics and cosmography. Munster's major work, *Cosmography*, was first published in Geneva in 1544: it was a massive work. The first edition had 659 pages with 520 woodcut maps and illustrations. Subsequent editions increased in size. The 1548 edition enlarged the work to 818 pages and 725 woodcuts. By the 1550 version, the work had reached gargantuan proportions of 1,233 pages and 910 woodcuts. It was published in all the major European languages as well as Latin. Thirty-six complete editions were published between 1544 and 1628. For much of the sixteenth century, it was the single most important source of geographical, historical and scientific knowledge. It was the great educational book of its era, an eclectic collection of material, some old, some new, part old myth, part new fact; the book contains material on surveying and mining techniques as well as discussions on goblins and spirits. It includes material detailing the reasons for sugar growing in parts of Italy as well as a discourse on the one-eyed and large-eared people who were supposed to inhabit parts of India. Discussions of latitude and longitude sit side by side with genealogies of long-dead European monarchies and Biblical exegesis.

John Dee (1527–1608) was one of the most interesting cosmographers of the sixteenth century.[2] Dee read widely and drew upon a large body of knowledge, some of which we would now call "occult." He was also concerned in affairs of state; he gave practical help to the imperial expansion of Elizabethan England. Scholar and public servant, savant and secret agent, Dee is an exotic character whose life and interests capture the zeitgeist of his age. He has not left a wide body of work. There is no cosmography that bears his name, yet he has had an enormous influence. He was part of a wider intellectual movement of Renaissance learning and Elizabethan statecraft. He had a huge library of cosmographical texts.[3] His library consisted of over 2,500 books at a time when the entire holdings of the University of Cambridge probably did not exceed 400 books. Dee traveled widely throughout Europe, including Italy, Austria, Bohemia and Poland. He returned to London in 1551 with an impressive network of contacts with leading cosmographers across Europe. He made frequent trips to Europe and corresponded with the two great map and atlas makers, Abraham Ortelius and Gerardus. He was very friendly with Pedro Nunez who was the Cosmographer Royal in Portugal. They discussed mathematics and navigation. When Dee was under detention for heresy, he made Nunez his literary executor.

The death of Elizabeth and the accession of James in 1603 mark a downturn in his relationship with the monarchy. James was obsessed with witchcraft; he feared plots against him based on occult practices and in the first year of his reign ordered Parliament to enact statutes to "uproot enchanters."

Dee, now friendless and powerless, was an easy target. In a petition to the king in 1604, Dee responds to the "slander" that he was a "conjurer or caller or invocator of divels." The Renaissance cosmographers became feared figures. Dr. Faustus and his diabolic contract become an evocative image. Giovanni Battista Porta was examined by the Inquisition in 1580, 1592 and 1598, and his work was consequently banned. The witchcraft craze is at its peak during Dee's later life. The great astronomer Kepler had to defend his mother against the charge, and in 1600 in Rome, Giordano Bruno was tried for heresy, excommunicated and burnt at the stake. A cloud of reaction is closing out the freedom of thought. Dee's final days are sad. His daughter had to sell off his books to get food and provide heating. He died in poverty in 1608.

Humboldtian science

The early cosmographers studied a wide horizon of human knowledge. Later, cosmography would splinter into astronomy and geography while astrology would join the nonscience category along with alchemy and natural magic. As knowledge expanded, it became more and more difficult to encompass the world, although part of the project of the Enlightenment was an all-encompassing understanding. The development of the dictionary and the introduction of the encyclopedia were attempts to capture the totality of language and knowledge, respectively. The Scientific Revolution involved a deepening, but also a narrowing of what was considered knowledge. Arguably, the last cosmographer in the old style is Alexander von Humboldt (1769–1859). From 1782 to 1792, he studied at universities in Frankfurt, Gottingen, Hamburg and Freiberg. Widely read, Humboldt had interests in economics, geology, botany and mining. He arrived in Paris shortly before the storming of the Bastille and, as he wrote later, it "stirred his soul." He had a lifelong social concern and radical political beliefs. In 1792, he joined the Prussian mining service where he devised better safety lamps and rescue apparatus. But there was also a mystical side; he always sought to identify the "life force" (vis vitalis). As the eighteenth century turned into the nineteenth century, Humboldt undertook expeditions to the Americas. He made careful measurements and was perhaps the first recognizably modern field scientist. His careful measurements of mountain ecology showed how altitude as well as latitude impacted biogeography. His careful measurements and sophisticated theorizing laid the basis for physical geography, plant geography and meteorology. He saw the environment as a laboratory that should be measured to reveal the workings of the world. He was also one of the first globally recognized scientific celebrities, fêted around the world. His name lives on in the names of counties, towns, streets and squares around the world. He had a tremendous influence on science and geographical studies. His influence is so strong that the term *Humboldtian science* is used to refer to precision in measurement, careful fieldwork and an awareness of social issues. Humboldt was a major figure in the development of nineteenth-century science. He corresponded with and influenced a range of writers and scientists including Charles Darwin and John Muir. The American poet Ralph Waldo Emerson described him as one of the wonders of the world. Humboldt's life work was *Cosmos*, published between 1845 and 1862. He originally thought of calling it *Cosmography*. It was a popular scientific book subtitled *Sketch of a Physical Description of the Universe*. He offered it to discern physical phenomena in their widest mutual connection, and to comprehend nature as a whole, animated and moved by inward forces. By the time he gets to a coverage of recent scientific theories in Volumes 3 and 4, the book is unwieldy, weighed down with references. The unity is collapsing under the weight of the material. *Cosmos* was one of the last great attempts to encompass the world. While Humboldt may be one of the first modern geographers, he was also one of the last cosmographers.[4]

Humboldt stimulated the mapping of environmental data. An 1823 global isothermal map by Woodbridge had the subtitle "Drawn from the accounts of Humboldt and others." Humboldt was enormously influential. He persuaded his friend, the German geographer, Heinrich Berghaus to produce the *Physikalischer Atlas* (1838–1848) as the graphical companion to *Cosmos*. It contained 60 maps of meteorology, climatology, hydrology, geology, magnetism, plant and animal distributions. In 1848, the Scottish cartographer A. K. Johnston produced a similar work entitled *Physical Atlas* that had full-page illustrations of original as well as Berghaus-informed maps. Figure 6.1 is a map from this atlas that draws on the work of Berghaus and Humboldt. The German cartographer August Petermann (1822–1878) was a collaborator with both Berghaus and Johnston, a sign of the transnational nature of the endeavor, and later went on to produce his own *Atlas of Physical Geography* in 1850. In France, Jean-Augustin Barral produced an atlas of physical geography under the Humboldtian-inspired title of *Atlas du Cosmos*. Humboldt laid the essential basis for the mappings of the physical environments in the modern national atlases of the late nineteenth and twentieth century.[5]

Even by the early twentieth century, a single cosmography could no longer embrace the widening world of knowledge. The torrent of information was overwhelming the ability to provide a single coherent framework. Yet, elements of the intellectual ambition of this wider arc remained. The national atlas of the twentieth century is a Humboldtian-informed version of applied cosmography. There is the same sense of encompassing a wide bandwidth of knowledge and the same attempt at comprehensiveness. It was a *Cosmos* for the nation-state. The nation-state was a smaller unit than the whole earth,

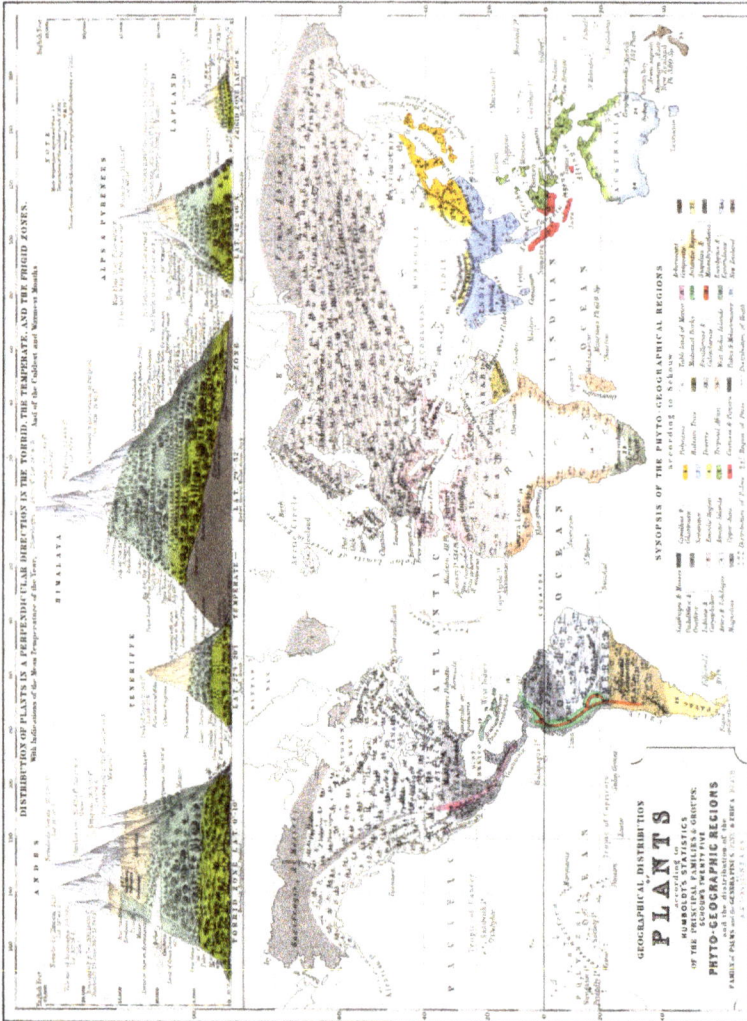

Figure 6.1 World map *Physical Atlas*, 1848. *Source:* https://en.wikipedia.org/wiki/Alexander_Keith_Johnston_(1804–1871)#/media/File:Geographical_Distribution_of_Plants_1850_Alexander_Keith_Johnston.png.

and so more amenable to that original promise of the earlier global cosmography. A line of shared, intellectual ambition links the early cosmographers of the sixteenth century with Humboldt of the nineteenth century and the national atlases of the twentieth century.

Science, State and the National Atlas

Knowledge is always collected for a purpose, for reasons related to power. The national atlas of the twentieth century is intimately connected to the state, governance and statecraft. We have already touched on some of the reasons of statecraft: the need for legitimacy, the need to create a resource inventory, providing an image of the nation-states to both citizens and a wider world, to be seen as modern and to be seen as a member of the international community of modern, independent states. Here, I want to focus on the national atlas as a coproduction of science and the state.[6]

What do we mean by "coproduction"? The national atlas is a product of a globalizing discourse of science. Scientific discourses as varied as climatology, demography, economics and ecology are employed and utilized in the national atlas. Similar themes appeared as the standards of the global epistemic community, including forms of data creation, data assembly and data display were widely adopted across the range of national atlases. The national atlas is a form of normal science where agreed-upon standards are followed and adopted. The atlas is both a product and a producer of science.

The range of topics covered and the way they are represented have a remarkable similarity in the many different national atlases as they all share a similar scientific viewpoint. Figures 6.2 and 6.3 are rainfall maps from, respectively, the 1928 *Atlas de l'indochine* and the 1979 *Atlas de Venezuela*. Although created in different political contexts, one colonial the other an independent nation-state, and at different times, over a fifty-year time span, they are very similar. Both employ a similar form of rainfall gradation and even color palette. They use standard hydrographic categories to create similar maps of the same phenomenon albeit of different places at different times. Both maps draw upon the same repertoire of scientific categorization and data display techniques.

The science in and of the national atlas is also intimately connected with the state and the state formation of scientific enterprises. The state is a central element in political thought and practice. General ideas range from the Marxist notion of the state as an embodiment of the interest of the capitalist class, to one that sees the state as having a measure of relative autonomy from the immediate interests of specific classes. Other viewpoints point to the progressive possibilities at times of crises for the state in capitalist societies. More

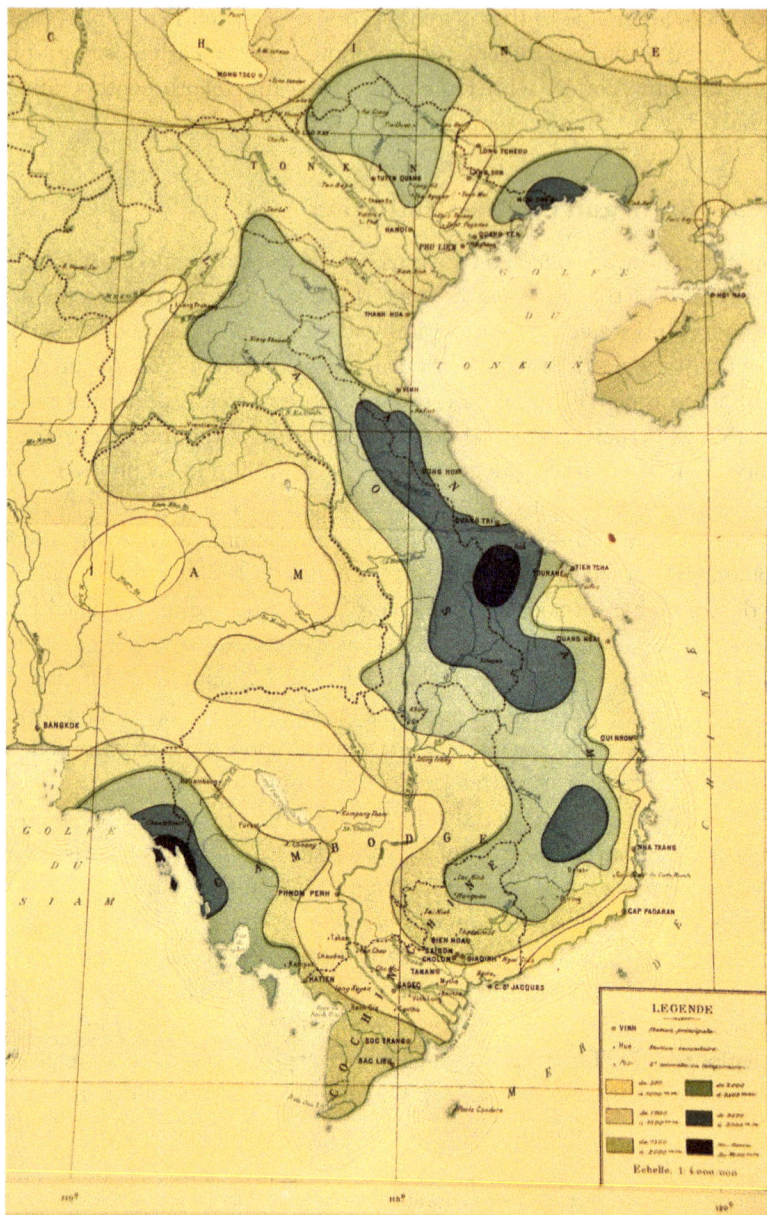

Figure 6.2 Rainfall map, *Atlas de L'Indochine*, 1928. *Source:* Photo by John Rennie Short.

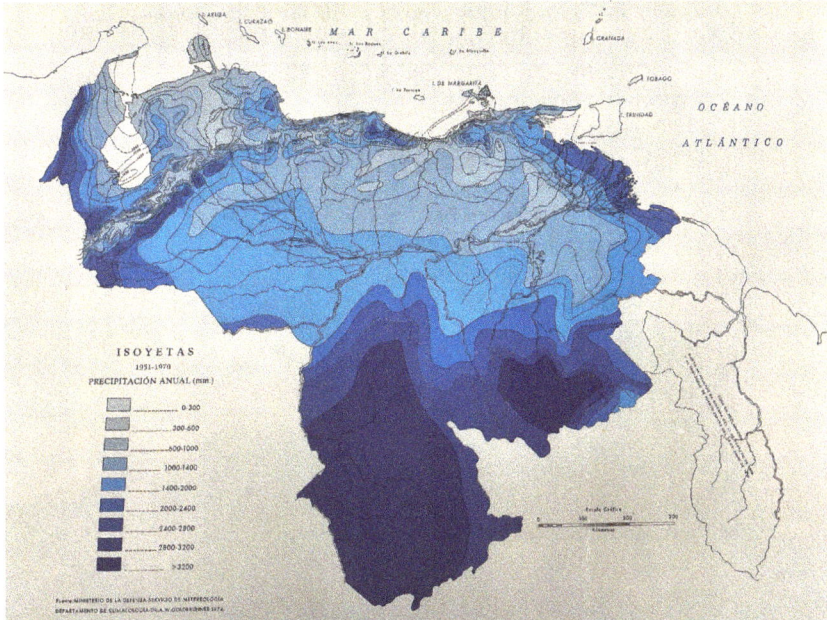

Figure 6.3 Rainfall map, *Atlas de Venezuela*, 1979. *Source:* Photo by John Rennie Short.

specifically, however, scholars have drawn attention to the role of science in state formation. States are constituted by knowledge practices just as knowledge is constituted by and through the state. The national atlas is a product of this relationship as the knowledge generation, production and dissemination of state practices are in turn guided and shaped by global and national epistemic communities.

Science is produced and reproduced by the state in and through the national atlas. The state, in the form of different government agencies, plays a vital role in providing data, expertise, modes of analysis, topics of investigation, modes of analysis and ways of thinking to the national atlas. Table 6.2 lists just some of the principal agencies connected with the production of the 1994 *National Economic Atlas of China*. Only the government and quasi-government agencies are listed in this table. Private companies and corporations were also involved. The atlases condense and facilitate a national science.

The notion of "scientific" was central to the idea of the modern national atlas. Arrays of experts were assembled, data were amassed and maps were produced. Data were turned into information and national narratives were mobilized to turn information and data into knowledge. Maps allowed data to be presented and explained in the context of the territory of the nation-state.

Table 6.2 Government Agencies involved in *National Economic Atlas of China* (1994).

Principal agencies
Office of Sate Council
State Committee of Family Planning
State Statistical Bureau
Institute of Geological and Mineralogical Information
State Committee of Science and Technology
State Committee of Education
Ministry of Public Health
State Committee of Sports and Education
People's Bank of China
General Customs
Centre of Surveying and Mapping Projects

Cooperating agencies
Chinese Academy of Social Sciences
National Bureau of Statistics
China Population Association
Family Planning Commission
Ministry of Heath
People's University of China
Peking University
China Coal

The atlas is an inventory of knowledge of the state, very often by and for the state. Science and power interact through and in the national atlas. The notion that the state can be viewed as an engine of science is expressed vividly in the *National Atlas of Korea* published in 2009. Table 6.3 lists the organizational members listed in the atlas. The national atlas represents the nation-state as a material ensemble of scientific practices. The preface to the 2009 *National Atlas of Korea* also lists an advisory board of 20 academics and scientists working mainly in universities and research organizations. The editorial boards of the different sections—national land and territory, nature and environment, population and settlement, economy and industry, social and political geography—consist of between 10 and 20 academic advisors. More than forty academics from a range of prestigious universities are also listed as contributors.

The territorial depiction of scientific data—climate zones, mountain ranges, river systems, etc.—renders the nation-state an objective and therefore legitimate entity. The scientific depiction is in part a claim to legitimacy through the deployment of the "objectivity" of science. The national alas is a coproduction of the state and science that buttresses the state's territorial claim and the objective reality of the nation-state.

Table 6.3 Government Agencies involved in *National Atlas of Korea* (2009).

Ministry of Land, Transport and Maritime affairs
Ministry of Strategy and Finance
Ministry of Education, Science and Technology
Ministry of Foreign Affairs and Trade
Ministry of National Defense
Ministry of Public Administration and Security
Ministry of Culture, Sports and Tourism
Ministry of Food, Agriculture, Forestry and Fisheries
Ministry for Health, Welfare and Family Affairs
Ministry of Environment
Ministry of Labor
Ministry of Gender Equality
National Intelligence Service
Korea Customs Service
Korean National Statistical Office
National Emergency Management Agency
Rural Development Administration
Korea Meteorological Administration
National Oceanographic Research Institute
National Geographic information Institute

National Imaginaries

The maps of the national atlas are not just simple depictions of an uncompli-cated external reality. The reality that they draw upon is not just out there waiting to be depicted. The national atlas does not so much capture the physi-cal spaces of the nation as much as it helps to create them. Their maps are a subtle and complex mix of information and national imaginings. I introduced the idea of national imaginaries in Chapter 1. They are, as I noted, ideas that connect space and society, identity and place, and meaning and territory in a series of active constructions and reconstructions. The distinction between the different types of discourses in the national atlas is hinted at in the pref-ace to the 1974 *National Atlas of Canada*. The editors noted that there were two types of discourses: one that depicted objective information and the other that was a contribution to, what they termed, "national self-awareness and cultural evolution," which is another way of describing national imaginaries.[7] The physical reality presented in the national atlas is, at one and the same time, not only part of a "scientific" discourse of what physical aspects should be shown, but also reflective of national imaginaries that provide a "national awareness." The physical depictions in the national atlas comprise not only observations but also longings, measurements as well as feelings, sentiments informed by facts as well as facts shaped by sentiment.

Arguably, the very first modern national atlas was the 1899 *Atlas öfver Finland*. It represented, in 32 maps of the country, a rural society. The general maps highlight a hierarchy of urban settlement all the way down to small villages. The churches of three different denominations—Lutheran, Catholic and Orthodox—are shown. There are no maps of urban areas. Finland was a predominantly rural society, but the atlas was not just reflecting an external reality, but also celebrating a folk culture that was the strongest expression of early Finnish nationalism. The folk culture was not merely expressed in the atlas, it was affirmed. To be sure, the country represented was a primary producer, mainly agricultural. There are maps of grain and wheat production and a colorful map of the export of sawmills to different countries of the world (see Figure 1.2). The 1910 edition of the atlas also has maps of the number of dairy cows. However, other imaginaries also appear. One series of maps stands out for their distinctiveness and level of detail, the maps of the telephone and telegraph lines. These do not appear in the more recent atlases. They have become so ubiquitous that they would pose great difficulties in highlighting telephone lines. Their presence in the earliest national atlas, which overall celebrates the agrarian and preindustrial, also highlights a very distinctive element of the national imaginary, Finland as a society held tight by modern communications. The land in the early atlases is a lauding of the folk culture that was such an important part of the cultural expression of Finnish identity and nationalism. But the national imaginary also contains ideas of a modernizing society. The Finns' love of telephones and communications is a centuries-old affair. The first telephone line was erected in Helsinki in 1877, barely two years after its first patent in the United States. Regional telephone companies were established across the country, and within fifty years, there were 815 companies. By the end of the twentieth century, over 95 percent of all Finnish households had a landline, and by 2007, 97 percent had at least one mobile phone. Finland has led the way in the diffusion of telephone lines and more recently in the production and adoption of mobile phones. An early mass producer of mobile phones was the Finnish company Nokia.

Two elements of Finnish national imaginaries—rural agricultural society and technologically advancing society—are thus highlighted and represented in these early editions of the national atlas. As the rural, agricultural imaginary weakens in the later editions of the national atlas, city maps appear, and the thematic maps now include levels of industrial production and manufacturing employment. The 1925 edition of the Finnish national atlas, for example, depicts levels of the "industrial population" across municipalities. By the

1976 edition, the stable, fixed world of small towns and villages was replaced with maps of commuting patterns across municipalities.

National imaginaries inform the national atlas in many ways: we can identify at least three. One is the notion of a modernizing nation. This is a particularly strong element in the national atlases of newly independent countries and those making a rupture with the past. The *Bol'shoĭ sovetskiĭ atlas mira* (*Great Soviet World Atlas*) is suffused with the national imaginary of a modernizing, urbanizing, industrializing society. The extent of electrification, for example, is a recurring theme of maps and tables. The growing industrial base of the country is not only recorded but also highlighted and celebrated. Figure 6.4, for example, depicts the distribution of different types of industries around Moscow. The sheer variety of symbols, indicative of a complex industrial base, makes the point. The design is clearly influenced by the lively Russian Constructivist movement and contains elements of the early art scene before the icy grip of Stalinist realism took complete control of Soviet art, design and aesthetics.

Second, there is the idea of a new society. In the case of Poland, for example, it was specifically the idea of a newly democratic society. It is not incidental that the first atlas of Poland printed after the fall of Communism, in 1993, the *Atlas Rzeczypospolitej Polskiej* (*Atlas of the Republic of Poland*) contains several maps of election results and party support. After years of political repression,

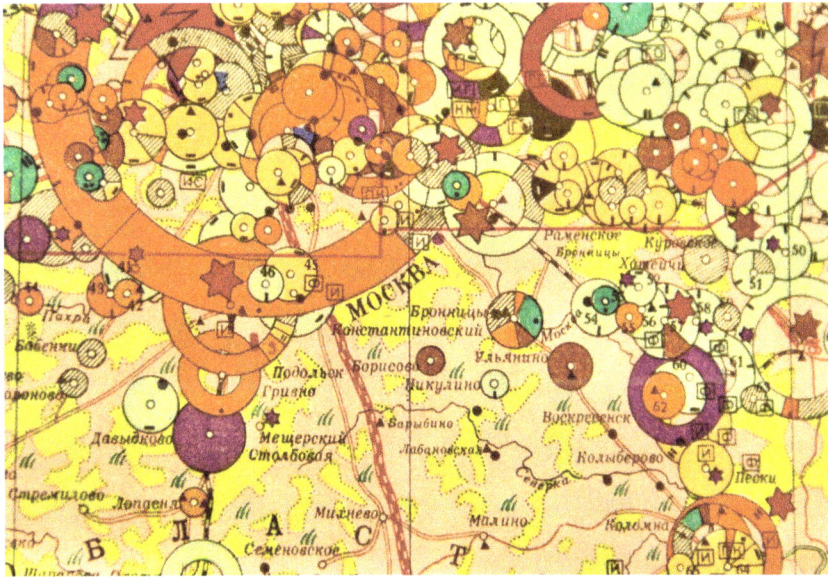

Figure 6.4 Economic activities in Moscow, *Bolshoi sovietski atlas mira* (*Great Soviet World Atlas*), 1939. *Source:* Photo by John Rennie Short.

the new atlas of the new democratic country chose election results as an important topic to be covered in maps and analysis. Figure 6.5 shows the support for one of the presidential candidates, Lech Walesa. It shows that the former Gdansk shipyard worker had a loyal basis of support in Gdansk and the more industrial and coal-mining parts of Poland. After years of authoritarian rule, the atlas celebrates the political geography of free elections.

Third, there is also the idea of the atlas as an important record of recent economic and social progress. The *Atlas of Iran White Revolution* was produced in 1973 by Iran's Ministry of Interior when the shah was in power and oil money was flowing in huge amounts allowing vast military expenditures as

Figure 6.5 Election results for presidential candidate, *Atlas Rzeczypospolitej Polskiej* (*Atlas of the Republic of Poland*), 1993. *Source:* Photo by John Rennie Short.

well as funding ambitious modernization projects. This atlas was used to highlight the achievements of shah's White Revolution. This revolution that occurred from 1960 to 1963 was a period of the consolidation of the shah's power that he used to promote land reform, establish literacy and health initiatives, marginalize and eliminate political opposition and reduce the power of the clergy. The celebratory atlas displays government policies of change and progress, including land reforms, literacy levels and forms of urban and rural reconstruction. Specific maps cover land reform measures, public ownership of forested land, increased rights for women, the formation of literacy corps to spread literacy in rural areas, the establishment of village courts and a whole panoply of urban and rural reconstructions. The emphasis is on the national space as the backdrop of social change and improvement. The atlas records the sweep of the shah's reforms and touts the success of the transformation,

> On the eve of the national celebration to commemorate the end of the first decade of the White Revolution, as the wave of inner happiness, joy and deep sense of gratitude spreads all over the country, this publication is issued to briefly portray the achievements of the past ten years.[8]

It was a revolution, however, that contained the seeds of the shah's overthrow. The revolution had a counter effect as it raised the ire of religious leaders. The "wave of happiness, joy and gratitude" mentioned in the atlas's text was not big enough or deep enough to save the shah's authoritarian and secular rule as the economy faltered and the rapid changes heightened anxiety and uncertainty. The religious fundamentalists launched an ongoing challenge that eventually led to the overthrow of the shah in 1978 and the establishment of a theocratic government in 1979, only six years after the publication of the celebratory atlas.

The national atlas is often used as a vehicle for representing recent social progress with the national space depicted as the scene of ongoing improvements and continued progress with progress in education and public health a recurring topic in what we may term these *atlases of modernization and social progress*. The *National Atlas of the United Arab Emirates* (1993), for example, contains prominent pictures of libraries and schools to show the progress of education. The 1999 *Atlas of Saudi Arabia* shows the increasing number of students graduating from the highly gendered education system. Botswana achieved independence from the UK in 1966. Like many other newly independent nations it set about the difficult task of providing basic government services to its population. The British provided little by the way of mass education. The provision of health and education became key indicators of progress. The

Botswana National Atlas, printed in 2001, noted in the preface that the national atlas provides a powerful glimpse of the country's efforts at development since Independence Day and was dedicated to the "building of an educated, informed, prosperous and proud nation."[9] Figure 6.6 shows a page from the large section devoted to education and training in the country.

As the imaginary changes, new maps are drawn, new topics are given cartographic priority and a new national atlas is produced. The earliest atlases of postrevolutionary Cuba, for example, are dominated by the rewriting of historical geography to highlight U.S. imperialist aggression. The later atlases

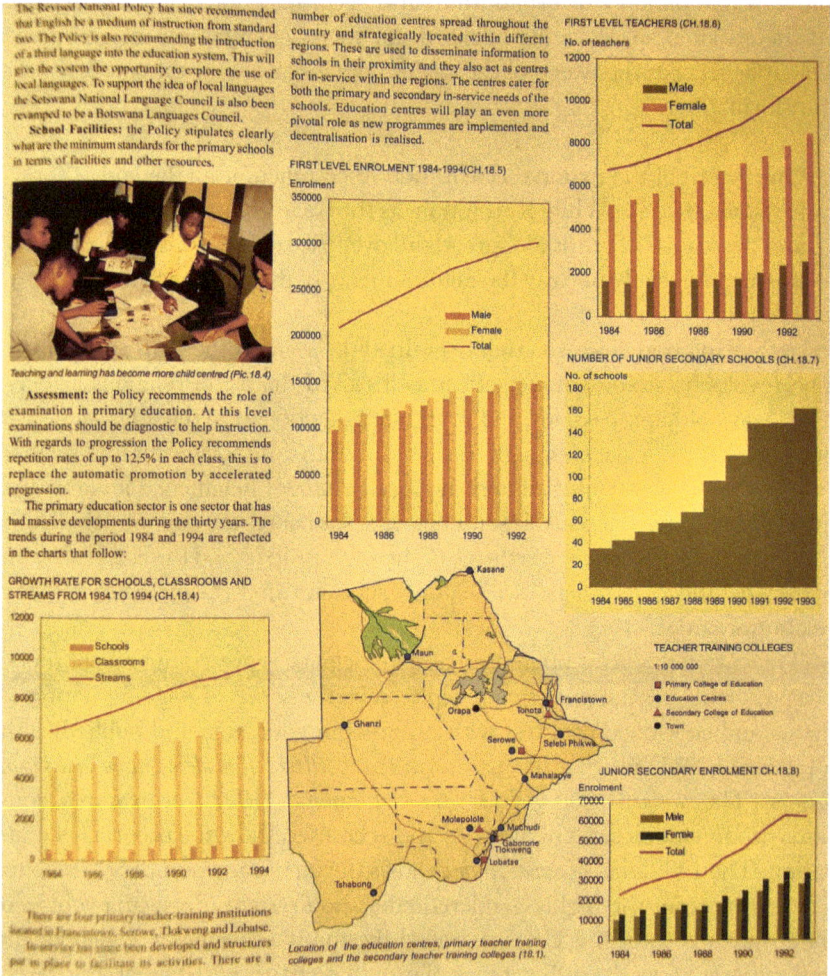

Figure 6.6 Educational progress in Botswana, *Botswana National Atlas*, 2001. *Source:* Photo by John Rennie Short.

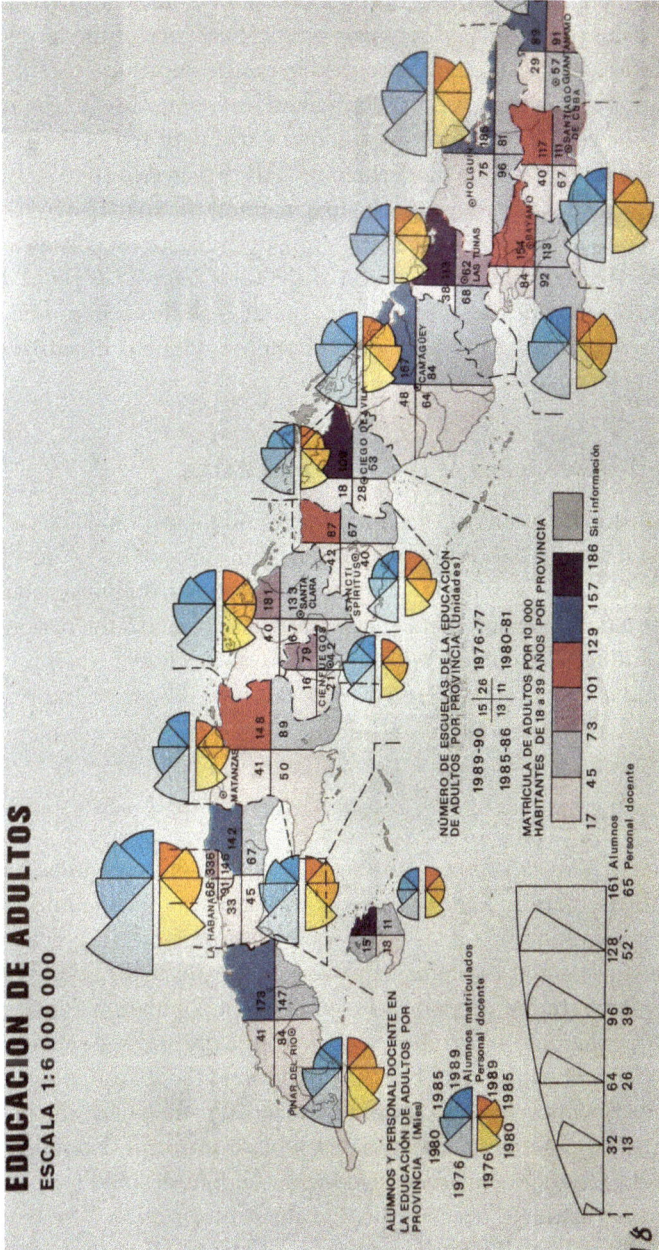

Figure 6.7 Educational progress in Cuba, *Atlas Socio-economico de Cuba*, 1991. *Source:* Photo by John Rennie Short.

dampen in revolutionary fervor and tend to concentrate more on social and economic progress, with maps and discussions of social progress made in education, public health and in the increase of women in the workforce. Figure 6.7 from the 1991 *Atlas socio-económico de Cuba*, one of the more technically sophisticated atlases, shows progress in adult education.

New national imaginaries are represented and embodied in the national atlas: from the established progress made in Cuba depicted in the 1991 *Atlas socio-económico de Cuba*, to the modernizing UAE depicted in the 1993 *National Atlas of the United Arab Emirates* and the first steps on the road to economic liberalization in the 1996 *Vietnam National Atlas*, to the more socially aware post-junta Brazil shown in the 2000 edition of the *Atlas Nacional do Brasil*. Despite the claim to objectivity and science, the national atlas also contains hopes and dreams, longings and regrets. It is the changing national imaginary given cartographic form.

Contested Discourses of the National Atlas

The relationship between national imaginary and the national atlas is neither fixed nor stable. Changes in public attitudes, displacements of government and paradigm shifts in science, all create sources of instability. Competing scientific practices and new ideological constructions make the national atlas a site of contested discourses. We can see this more clearly when we look at the evolution of what topics were covered and how they were framed in different editions of the same national atlas.

Changing paradigms

The 1966 *Atlas Nacional do Brasil*, produced under the military junta, has a traditional subject matter. The major sections include politico-administrative, physical, demographic, economic and sociocultural. In the physical section, Brazil was represented by maps of geology, climate, hydrography and climate. The demographic section included maps of population density, urban and rural populations, age groups, the economically active populations and the colonization of Amazonia. The 2000 edition of the *Atlas Nacional do Brasil* was produced by more critical social scientists after the fall of the junta and the introduction of a more democratic form of government. Even the categories adopted to frame the map changed with sections devoted to "configurations," "restructurings," "decentralization and centralization," "restructuring of agrarian space" and "the urban question." Physical maps in the post-junta atlas now show biodiversity and deforestation, while the demographic maps include infant mortality and health levels, topics that did not even appear in the earlier editions. The type of information that was mapped changed not simply because of improvement in understanding but because of an epistemological break in what is considered important types of space to be mapped.

The reimagining of physical space is even apparent in the different national atlases of Brazil. The 1966 atlas, produced during the military regime, depicts the Amazonia interior as a place to be developed. One map proudly shows the movement of people, indicative of the wilderness being transformed by resilient Brazilians. The 2000 atlas, produced by a more democratic and left-leaning government, depicts areas of biodiversity, and the Amazon is depicted in maps as a contested vulnerable space with assassinations of workers and peasants. Figure 6.8 shows one of a series of maps that plots the assassination of fieldworkers in five-year intervals. The Amazon is no longer an empty space to be filled by capital flows and settlers. It is a site of environmental concern and social conflict.

A similar type of epistemological break is also evident when comparing the colonial with the postcolonial atlas. Take the case of Ghana. In the colonial *Atlas of the Gold Coast* when it was still known as the Gold Coast, produced in 1928 and 1945, only certain forms of economic space are depicted.

Figure 6.8 Assassinations of fieldworkers in Brazil, *Atlas Nacional do Brazil*, 2000. *Source:* Photo by John Rennie Short.

British colonial authorities were primarily interested in the Gold Coast as a colonial plantation. The two colonial atlases depicted the economic space of agricultural products of the export market, including cocoa, coconuts and palm oil. These commodities were owned by foreign companies with little positive economic spillover effect to the local economies. They were part of the economic exploitation of the land for primary export products for the world market. Only the economic spaces of export-oriented produce were mapped, as these were the primary concern of the colonial authorities. The British were in Ghana not to develop the region or to enrich the local people; they were there to ensure cheap products for the British market and profits for corporate interest. In contrast, the 1970 *National Atlas of Ghana* was produced after independence in 1957. Notice how the title describes the national project of a new nation with a new name. The postcolonial authorities were now concerned not only with the export markets, still a vital source of foreign exchange, but also with the economic lives of most of the Ghanaian people. The maps of economic space now show not only the export crops but also the economies of everyday lives, including maps of the distribution of pig, poultry, sheep, cattle and fishing. The economy was imagined not just as an export site for foreign markets but as the activity space of most of the population. The 1970 postcolonial atlas of Ghana was more of a national economic inventory than simply a depiction of a primary exporting economy.

The postcolonial atlas also involved a change in how physical space is classified. The 1959 *Atlas of Kenya*, produced under British colonial control, used the landholding categories of Crown land, native land and Highlands. This categorization reflects the alienation of native land and its allocation to the government and settlers. The first category of Crown land is government land, taken over from indigenous owners. The third is private land allocated to and held by (only) white settlers in the Highlands. The second category of native lands is further subdivided into native land unit, native settlement area, temporary native reserve, native reserve and native leasehold land. In the postindependent *National Atlas of Kenya* of 1970 and 2003, the term "native land" is no longer used and a new categorization of government land, "trust land" and "private land," is employed. The shift from colonial to postcolonial is marked in the atlases by the deracialization of the landholding categories. The physical spaces are not just renamed: they are reimagined. Under the British, it was important to identify government holdings, areas of land reserved for white settlers and the range of native lands used to manage, control and contain the "native" population. The postindependent government no longer needs to make the same distinction. The goals have changed. The colonial eye saw native lands as a cause for concern in the management of a subjugated people who were showing massive resistance to white rule. The postcolonial government saw the land categorization of the atlas as a reference material for education and research, to provide a national resource inventory and to project national cohesion. Even urban space was imagined differently. In the colonial atlases of Kenya, the main buildings recognized in the town maps were golf courses, racecourses, police stations, prisons, hospitals and schools. In effect it was a mapping of the infrastructure of white colonialism, the whites in power and at play. The names of the dwellings of the local people are described with the simple and generic category of "African Locations."

When we compare the atlases of former French colonies with those of the postcolonial nations, a similar shift occurs in how space is viewed and what is considered important and what is depicted. The 1925 *Atlas des Cercles de L'A.O.F* depiction of Senegal is concerned with administrative boundaries, ethnic differences, roads and railways. The infrastructure of communication is vital for an occupying colonial power. Roads are categorized by ease of movement. Telegraph stations, post offices and customs posts are highlighted. The emphasis is on physical space as an administrative issue of effective control and communications. This was an element also apparent in the French mapping of Indochina (see Figure 4.9). The 1977 national atlas of an independent Senegal, *Atlas National du Sénégal*, has a much broader conception of what physical spaces are important. There are, for example, maps of intra-country movement. This was important information for the authorities dealing with the consequences of the massive rural to urban migration. The city maps highlight informal settlements. The postcolonial atlas is trying to capture the new spaces of rapid, unrivalled, urban growth. Yet in a reminder of the continuing colonial entanglements, the atlas of the independent country was published in Paris by the French National Geographical Institute.

The *Vietnam National Atlas* was produced in 1996 as a panoptic view of the territory to mark a new era for the country. The aim is to show a country under active reconstruction. But it is not a white washing of difficulties. The text highlights problems of the labor force, lack of diversity in labor arrangements and manpower planning. It is critical of the thin spread of capital investment, of poor-quality production and of a seriously defective system of subsidization. It concludes that the road to achieving a comprehensive and powerful industry is still a long one. The atlas is probably one of the more critical public documents to emerge from Vietnam of that era, a very critical assessment of the efforts to modernize and industrialize.

Delimiting boundaries

National space is bounded. The boundaries used to depict the nation-state are not only statements of fact but also longings and arguments. They are a geographical fact but also can be a remembrance as sometimes a form of forgetting. The successive editions of the national atlas of Finland allow us to witness changing attitudes to state and territory. The early editions and especially the 1899 edition were produced when Finland was ostensibly under Russian control. Finnish territory and identity were brought together and crystalized by the atlas. The 1925 edition consolidated the newly independent state. The first editions immediately after 1940 had to cope with territorial loss to the USSR during the war. Following defeat by the USSR, Finland was forced to cede the eastern province of Karelia. Stalin then ordered the deportation of the Finnish speakers. The less populated territories of Sall and Kuusamo, along the eastern border, were also ceded. More than 11 percent of the national territory was ultimately ceded to the Soviet Union. The 1960 *Atlas öfver Finland* has a series of maps that show the changing boundaries of Finland between 1751 and 1947. While the 1960 atlas shows a series of maps of territorial loss, this does not appear as a major theme in subsequent national atlases. The loss of territory has been accepted—there is no national imagining of a contested territory or of taking back territory. This is not so with other countries.

The *Atlas de la República Argentina*, in all its various editions since 1886, depicts the islands off its southeastern coast as *Las Malvinas*. South Georgia, long internationally recognized as belonging to the UK, is also represented as part of Argentinian territory. To make the position crystal clear, the 1954 edition uses the designation *(Arg)* after the names of the two sets of islands to reinforce the claim. There is no indication from all the different editions of this national atlas that the islands are de facto and de jure part of the UK. The atlas also ignores the UK claim to Antarctica, depicting all the British claim as Argentinian territory. The territorial depiction of the country does not accord with the internationally agreed jurisdiction. The atlas is as much about longing and national pride as about geopolitical realities.

In a similar fashion every thematic and the relevant regional and all national topographic maps in the 1979 *Atlas de Venezuela* continue to press territorial claims. The 1979 atlas has, in many respects, a modernizing, progressive perspective looking to a bright future of economic development based on rich oil reserves and expanding manufacturing base. Yet it also has echoes of the 1840 atlas which depicted "territory usurped by the English" (see Figure 3.1). The 1979 atlas has the wording "Area in relation to the Geneva Agreement of February 17, 1966 and the Spanish port protocol of June 18, 1970" across all the thematic maps (see Figure 6.3). While the language may be slightly different the sentiment is the same. This territory really belongs to Venezuela.

All the modern national atlases of China, for example, show Taiwan as part of China and show a giant lobe of water in the South China Sea as part of China's sovereign territory (see Figure 5.2). The vast claim impinges on the exclusive economic zones of Vietnam, Philippines and Malaysia and is not recognized by the international community. The cartographic claims remained little noted until a more expansionist and muscular China emerged under the leadership of President Xi in 2013. The maps were merely claims on a map until a stronger and more assertive China turned these claims into reality as islands in the South China Sea were appropriated and militarized. The maps were "promissory notes" that came to fruition when China's military strength allowed them to be acted upon.

Pakistan's national atlases continue to lay cartographic claim to Jammu Kashmir in maps that show them in the same-colored depictions as the rest of Pakistan. Indian territory is shown as a sparse space, a dull space with few features. Kashmir, in contrast, is represented in the rich diversity of colors and features, the same as Pakistan territory. Kashmir is depicted as if it were an organic part of Pakistan and separate from the bleak mass of India. In other maps in the 1997 *Atlas of Pakistan*, Kashmir has red lettering that describes it as disputed territory (see Figures 3.5 and 5.4).

Sometimes the national atlas is a record of a more expansive state. The 1965 *Atlas Nacional de España* contains the remnants of Spain's once vast empire. The atlas has a section entitled *Plazas y Provincias Africanas* that contains maps of Spanish Sahara, Spanish Guinea, Fernando Po and Ifni. These were all part of Spain's African territories. The atlas was produced when Spain still held substantial territory in western Africa and just before decolonization. Subsequently, Spanish Guinea became the independent country of Equatorial Guinea in 1968, soon joined by the island of Fernando Po. In 1969, Ifni became part of Morocco. Spanish Sahara, now Western Sahara, remains problematic: it was ruled by Spain from 1884 until 1976 when Spain gave up its claim. When Morocco claimed the territory, a resistance movement of the local Sahrawi peoples led to a war that lasted until a cease-fire in 1992. The UN considers it a self-governing territory. In 2020, under the Trump administration, the United States formally recognized Morocco's claim to the territory.

At other times, the national atlas can even "forget" the land under state control. The 1970 *National Atlas of the United States*, for example, contains hundreds of thematic maps covering topics from agricultural production to urban areas. But most of these maps only contain some of the national territory, the continental United States, Alaska and the Hawaiian Islands. There is a small section of the atlas entitled *Outlying Areas of the United States* which lists areas under the sovereignty of the United States. These include, in the Pacific, American Samoa, Guam and the trust territories of the Carolinas, Marianas and Marshall Islands. In the Caribbean, the areas include the U.S. Virgin

Islands and Puerto Rico. These are all under U.S. authority and control and yet are not part of the "national" thematic maps. They inhabit a penumbral category of being American yet not American territory. In a fascinating book Daniel Immerwahr argues America was such a late imperial power with a national rhetoric of promoting democracy that it consciously hid its empire.[10] The empire of the "greater United States" was hidden from public view. The term "outlying areas" used in the 1970 atlas sounded so much better than "colonial territories." The 1970 national atlas, purportedly an atlas of the national territory, is coyly silent throughout most of the massive text about the existence of these overseas territories. While mentioned in a very small specific section, they are not accorded the dignity of cartographic presentation in the over 700 "national" maps. The national atlas does not depict the greater United States and hides some of the nation's territory.

Politics of toponymy

In 2009, I was examining national atlases in the AGS cartographic collection at the University of Wisconsin, Milwaukee. I came across the 1954 *Atlas de la República Argentina* produced soon after the 1952 death of Eva Peron, wife of Juan Peron, the Argentinian president from 1946 to 1955. On close examination, I noticed that the atlas had been marked. On page 13 in the title and on the map, the name of the province *Eva Peron* was scored out in red ink and replaced with *La Pampa*. Similarly, in a later page the name of a province *President Peron* was replaced with the name *Chaco*. On a map of cities, the name *Eva Peron* was similarly defaced and replaced with *La Plata* (see Figure 6.9). The defacement was more official than unofficial. I checked the records and found that the atlas was a gift of the publisher, the Institute of Military Geography and Ministry of Education, in 1956. So this was not a random act of vandalism but an official if crude renaming. The Peron name did not survive the Peronist regime and to this day the provinces are still known as Chaco and La Pampa and the city on the coast of the wide estuary of River Plata is still La Plata.

The names used on atlases can be erased, defaced and forgotten. They can also be contested. Naming is an act of power that privileges some name-giving groups over others. The condescension of history erases some names, while others remain contested. Let's look at one example of name contestation in national atlases in more detail.

East Sea/Sea of Japan

In the English versions of the 2009 *National Atlas of Korea* and the 1977 and 1990 *National Atlas of Japan*, the shared sea between them has two very

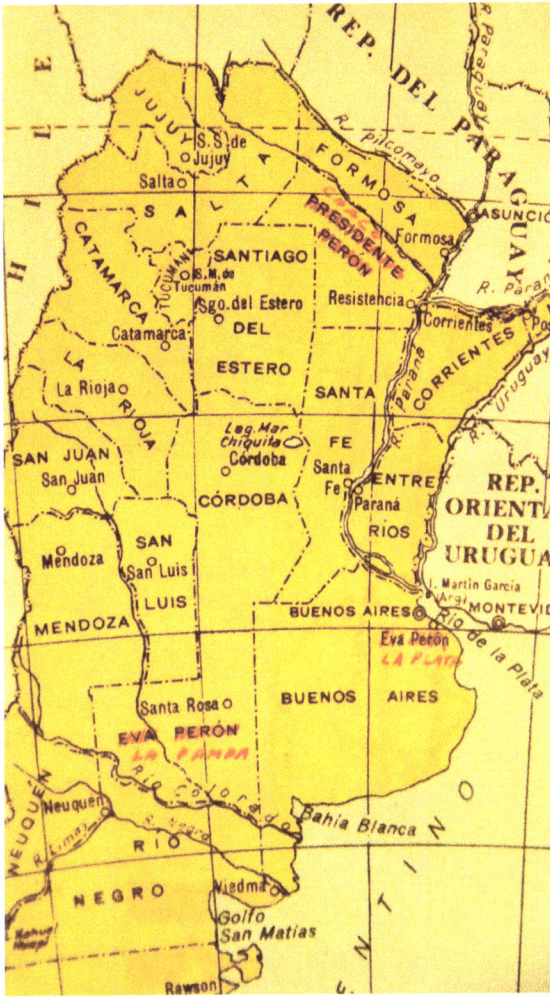

Figure 6.9 Renamings, *Atlas de la República Argentina*, 1954. *Source:* Photo by John Rennie Short.

different names. In the Korean atlas it is referred to as East Sea while the Japanese atlas uses the term Japan Sea (see Figures 5.8 and 6.10). Behind this different naming lies a complex history.[11]

There is a very long history in Korea, dating back at least 1,500 years of the term "East Sea" being used to describe the body of water to the east of the Korean peninsula. By the eighteenth and nineteenth centuries, Western explorers used numerous names including Sea of Corea and Gulf of Corea. Even early Japanese maps sometimes used the term Sea of Joseon or Sea of Japan to reference the same body of water.

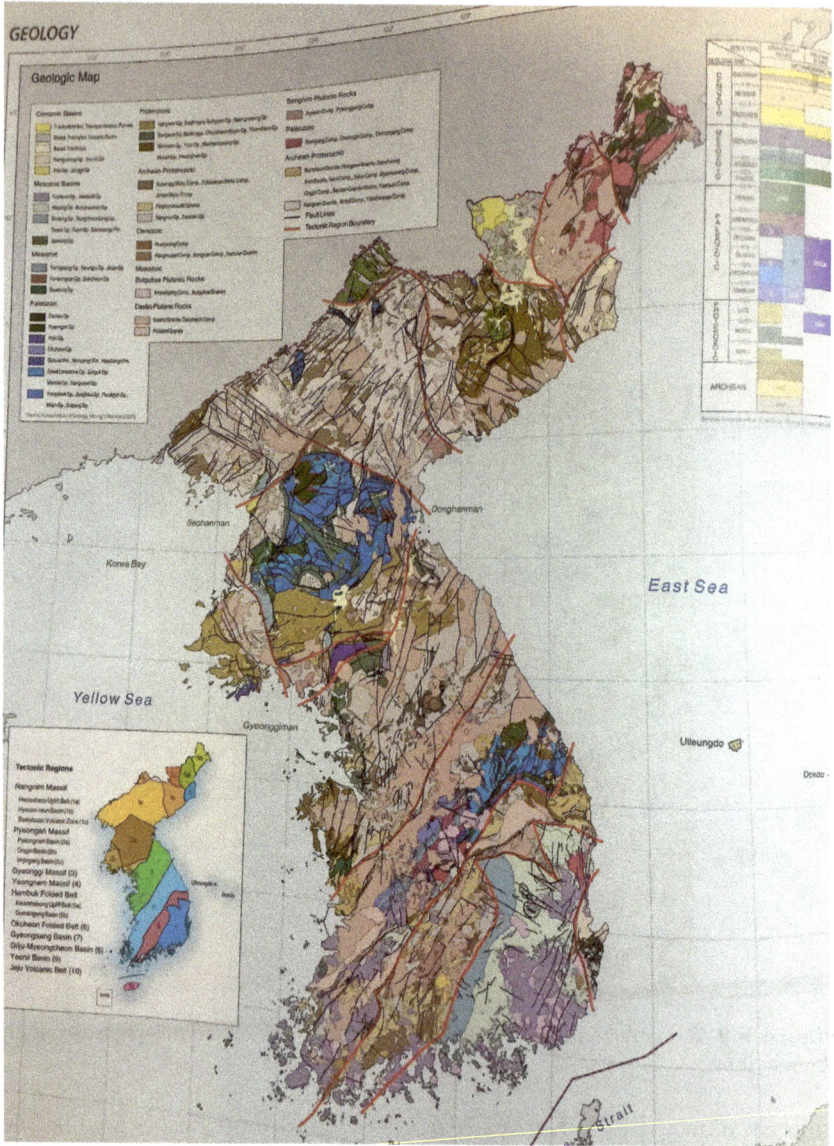

Figure 6.10 Geology map, *National Atlas of Korea*, 2009. *Source:* Photo by John Rennie Short.

As part of an expansive Japanese imperialism, Korea became a Japanese protectorate in 1905 and was annexed formally in 1910. The 1943 *Atlas of Japan*, produced by U.S. Naval Intelligence from Japanese sources, depicts Japanese national territory as including not only Japan but also Korea and Taiwan. The domination took many forms, including a toponymic colonialism. A Japanese naming system was applied to all maps, including place names, roads and natural features of the country. The renaming was a toponymic expression of Japanese power. There was also the renaming of larger regional features, such as the Sea of Japan, envisioned now as part of a wider Japanese Empire. The Korean designation of "East Sea" was erased from official maps and other formal documents.

The singular use of the name "Sea of Japan"—with its erasure of "East Sea"—was formalized in international usage at the 1929 Conference of the International Hydrographic Organization (IHO) when the term "Sea of Japan" was recognized as the only official name and reconfirmed at subsequent meetings of the IHO in 1937 and 1953. Korea had no independent voice at the initial meeting, and there was little international sensitivity to the colonial context of the new name. Even after Japan's defeat in World War II, there was little formal reckoning of Japan's colonial rule in Korea. The occupying U.S. forces used Japanese maps of the country, and the name "Sea of Japan" became the widely accepted name. The colonial underpinnings of the designation "Sea of Japan" were neither addressed nor considered. Some of the South Korean military and economic elites prospered under the period of Japanese control and had little interest in raising issues about South Korea's colonial relations with Japan. However, 1960 marks a change. The brief Second Republic (1960–61) saw the beginnings of a more critical perspective of Japan. Military rule (1961–63) and then a series of authoritarian governments (1963–87) adopted a more overt nationalism and East Sea became a commonly used term. Then, as South Korea democratized, there was an even greater articulation of the colonial experience's negative aspects. The name "East Sea" embodied this nationalist discourse drawing attention to a separate Korean identity before Japanese occupation. Yet, South Korea remained insular, exerting little influence in the global-naming discourse. The series of authoritarian leaders and military dictatorships reinforced this sense of isolation. Disconnected from the international community, and especially from the international regimes of naming practices, South Korea had few opportunities to make its case on the international stage. In September 1991, both South and North Korea joined the United Nations. A more globalized South Korea could now address the naming issue in international forums. The official Japanese position is to resist either changing the designation of Sea of Japan for East Sea or to adopt a dual name. They argue that "Sea of Japan" is widely accepted in international usage and adopting a dual name would be confusing.

In Korea there is now a powerful constituency that sees the name "East Sea" as a central feature of national identity. In recognition of this constituency, all South Korean official government documents use the term "East

Sea." Korean agencies have been successful in promoting the use of the name "East Sea" in international newspapers, textbooks and other forms of communication. While Korea and Japan continue to use their own variants in their respective national atlases and other official maps, the dual naming of East Sea/Sea of Japan is becoming more common across the world.

The national atlas in the Anthropocene

There was a change over the twentieth century in the way that the physical space of the nation was represented in the national atlas. In the first two-thirds of the twentieth century, the territorial space was imagined almost as a lifeless container, at best a place with economic potential as a resource base, the source of economic growth and development. The maps of oil refinery capacity from the 1999 *Atlas of the Country of Saudi Arabia* (Figure 6.11) or of crop production in the *1970 Atlas nacional de Cuba* (Figure 6.12) are typical examples that portray the environment as the backdrop for economic activity. By the last third of the twentieth century a new view of physical space emerged, often given the name the *Anthropocene* in which the environment was

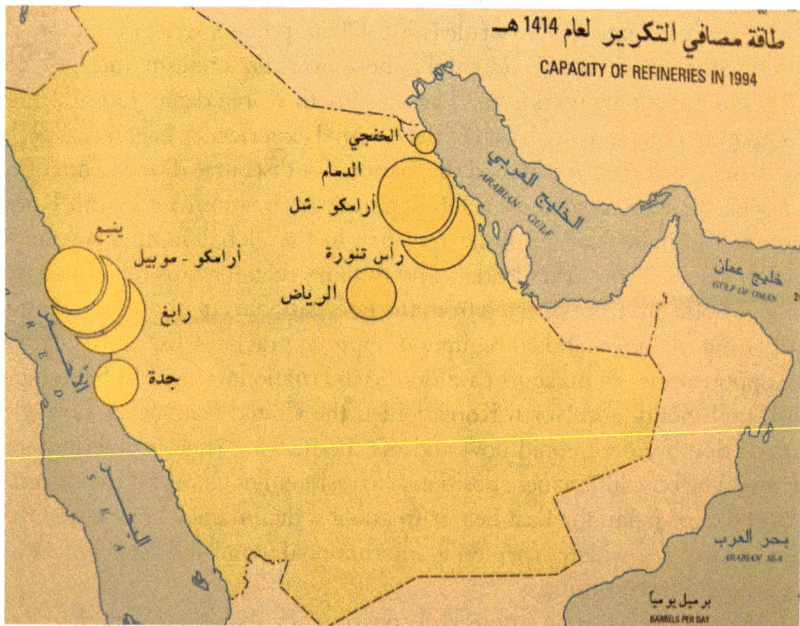

Figure 6.11 Oil refinery capacity, *Atlas of the country of Saudi Arabia*, 1999. *Source:* Photo by John Rennie Short.

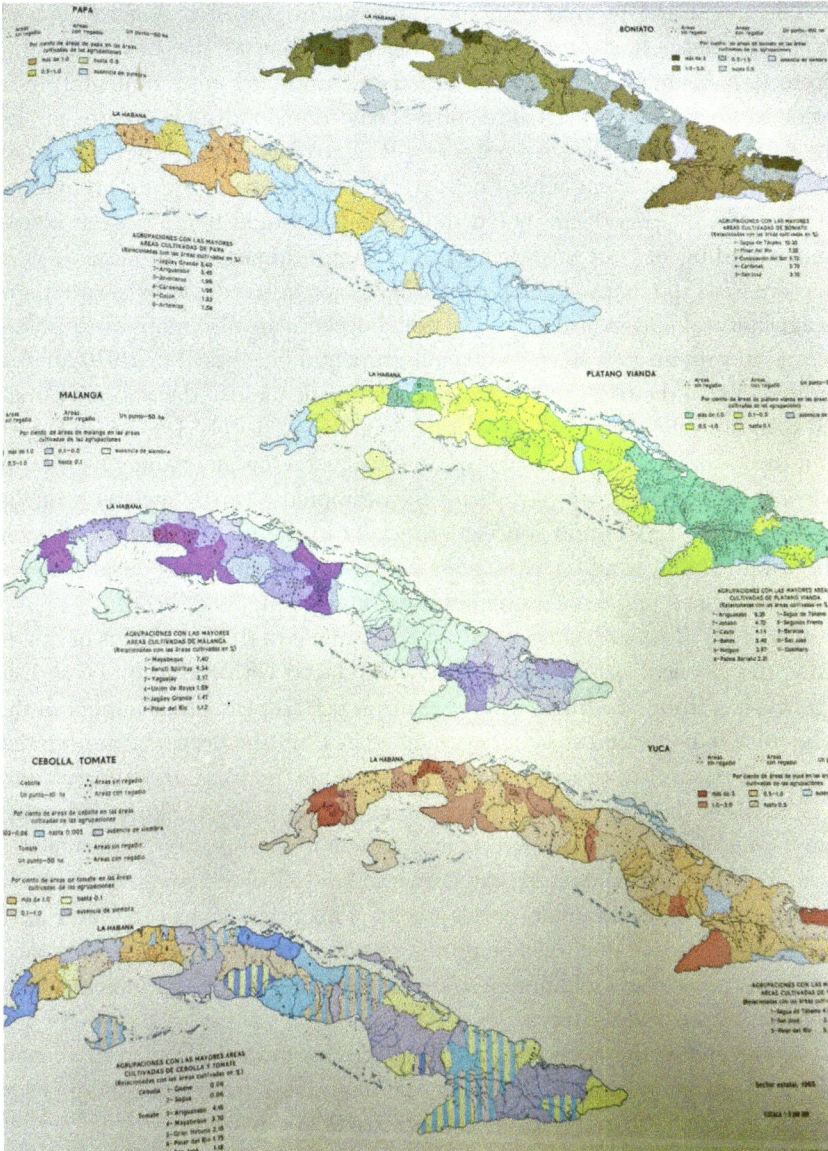

Figure 6.12 Crop production, *Atlas nacional de Cuba*, 1970. *Source:* Photo by John Rennie Short.

seen as a living entity, a vital organism that needed to be managed and protected.[12] There was a paradigm shift from the view of national physical space as an inert container to a living organism and from a focus on commodities to sustainability.

The beginning of change is most obvious in the 1970 *National Atlas of the United States*. It was produced at the same time as the Environmental Protection Agency was being established and landmark environmental legislation was being introduced. The spirit of environmentalism was in the air. In the dedication page, the then president, Richard Nixon, perhaps best known for Watergate but along with Theodore Roosevelt one of the most environmental of U.S. presidents, noted that the atlas would help create a better understanding of "our environment and man's impact on it." The atlas is massive with 765 maps. They are the usual economistic depictions with maps of agricultural and industrial production, but there are also maps of air pollution, water impurities, storms and environmental hazards. This 1970 atlas is on the cusp of the older traditionalist economistic view of the nation's physical space and the emerging one of a national space that needs to be managed and cared for as much as exploited and utilized. Climate change has yet to become an issue. So, we have maps of traditional hazards such as drought and flooding, a perennial concern especially in the interior but nothing on the wildfires that would only escalate in size and number and duration with the pronounced impact of a warmer, drier U.S. West.

The 2001 *Botswana National Atlas* is an example of the new forms of imagining the physical space of the country. Produced with the help of Swedish agency, it is imbued with forms of Nordic and European environmentalism. The preface notes that since independence in 1966 the need to conserve the country's natural resources has been increasingly realized and appreciated. The maps in the atlas are grouped around themes such as culture and economy. One of the main themes is environmental conservation with maps of forest reserves, national parks, game reserves and wildlife management areas. The focus is on sustainable development. This atlas depicts the land as a renewable resource that must be protected and nurtured. The national space is not just a site of exploitation, a source of commodities, but as a living thing to be protected and nurtured.

By the end of the twentieth century, attitudes toward physical space had changed. At the century's beginning, the national space was a container for resources that were to be used and exploited. The maps of geography that feature in all the atlases were visualized as backdrop to utilizing the mineral deposits and the weather and soil maps providing the necessary data for effective agriculture. But by the end of the century the environmental crisis was more obvious. Economic growth, which had long been viewed as the ultimate national goal and the main purpose of many of the thematic maps in the national atlas, was questioned. Environmental deterioration was more apparent as the weight of increased technology impacts and growing population pressure tested the livability of the planet. The national atlases began

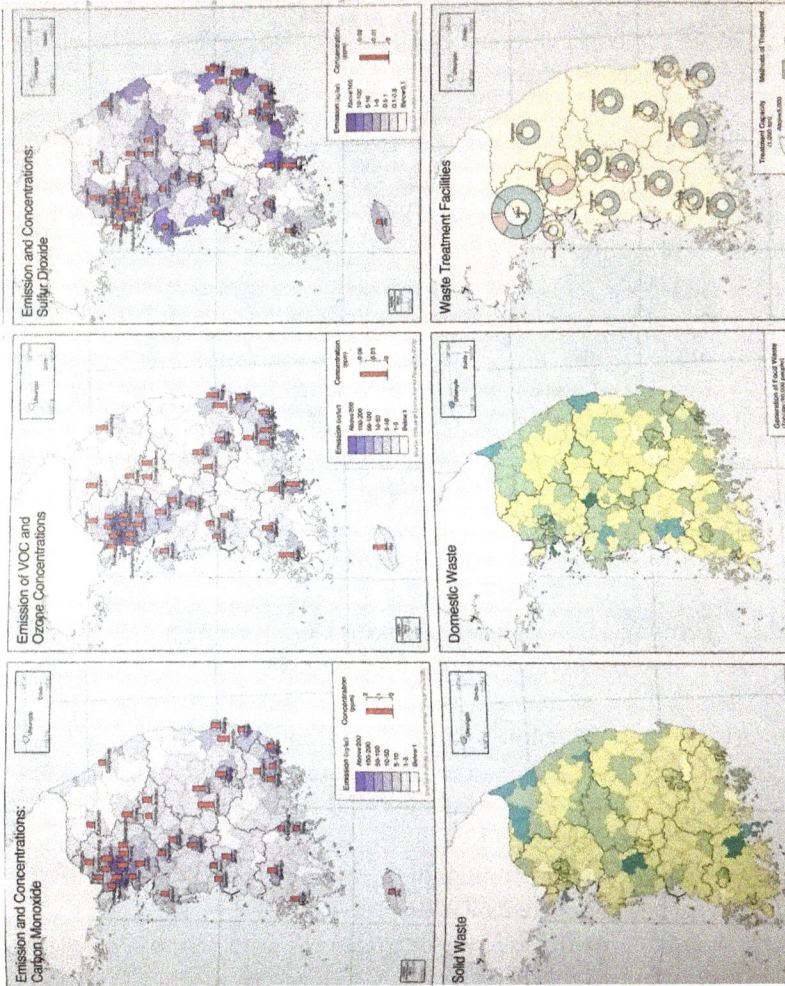

Figure 6.13 Pollution maps, *National Atlas of Korea*, 2009. *Source:* Photo by John Rennie Short.

to reference this change in mood. The notion of the environment as inert, lifeless matter, the backdrop for economic exploitation was more and more replaced by the sense of a living planet challenged by environmental deterioration. Figures 6.10 and 6.13 both come from the 2009 *National Atlas of Korea*. While some variant of Figure 6.10 can be found in almost any national atlas of the twentieth century, Figure 6.13 with its maps of pollution levels has a distinct late twentieth-century sensibility.

Consider the national atlases of China. In the preface to the 1987 *Population Atlas of China*, the editors noted that the atlas was a necessary tool for China's drive toward socialist modernization. The 1994 *National Economic Atlas of China* has a traditional suite of 265 maps. The maps of geology and vegetation, for example, are typical of the national atlas maps of the twentieth century and similar in form to Figure 6.8. The overall emphasis is on the physical environment as the silent setting for economic growth, industrial development and rising standards of living. This 1994 atlas depicts spaces of production and revels in China's economic success and the level of recent growth. There are maps of textiles manufacturing, machinery production and ferrous and nonferrous mining. Later national atlases, however, articulate a more critical approach highlighting some of the costs of economic growth. The *National Physical Atlas of China*, published in 1999, has maps of environmental hazards, including dustfall in cities, the frequency of acid rain, tidal surges, cold snaps, debris slides, desertification and soil erosion. The tone is even more critical in the 2000 *Atlas of Population, Environment and Sustainable Development of China*. The title is revealing. Now there are maps of endangered species and a map of ecosystems ranked on environmental quality. The atlas is concerned with the ecological and health implications of rapid economic growth. The Foreword notes that

at the dawn of the 21st century, China is confronted with heavy population and environmental pressures. [...] China has put poverty alleviation and illiteracy education at the top of the agenda. Our ecological situation is also at a critical stage: soil erosion is serious; lakes are shrinking; grasslands are deteriorating; forest coverage is low; and rare and endangered species are disappearing. The rapid industrialization and high population growth have created great pressure upon the environment. The air in two-third of our cities is seriously polluted [...] surface groundwaters are polluted. [...] Lakes are polluted. Acid rain is serious [...] environmental pollution is constraining to the economic development, and it poses a serious health threat.[13]

The physical space of China is now understood and represented not just as the platform for economic growth and modernization but as a damaged and vulnerable resource. The maps now revolve around issues of sustained development, protection of the environment and controlling pollution.

We can also see the change when we focus in on a specific topic. A study of soil maps in atlases, for example, is revealing.[14] Almost 80 percent of national atlases contain such maps. By the middle of the twentieth century, most national atlases contained one or more soil maps focused almost entirely on soil classification. They showed the different types of natural soil distributed across the national territory. Soil was considered a resource classified by its usefulness for agriculture. But by the end of the century and into the twenty-first century, the soil map section dealt with anthropogenic issues such as soil conservation, soil degradation, soil erosion, soil pollution, soil resilience and chemical pollution of soils.

The national atlas in the age of the Anthropocene portrayed the environment as both a product of anthropogenic activity and as an invaluable resource that needs careful management and protection to ensure sustainability. This trend is evident in the emergence of pollution as an important topic covered in national atlases. For example, compared to earlier editions of the national atlas of Israel, the 1985 *Atlas of Israel* and the 2011 *New Atlas of Israel* have maps of environmental quality, waste disposal landfill sites, wastewater treatment plants, noise pollution from Ben Gurion airport and levels of nitric acid and sulfur dioxide. In many other national atlases, spaces of pollution are now considered legitimate topics for cartographic consideration. The 1993 *Atlas Rzeczypospolitej Polskiej* (*Atlas of the Republic of Poland*), for example, is the first Polish national atlas to discuss land degradation and pollution, items not considered in the atlases of the communist era.

Physical space is also reimagined as a source of hazards and disasters. The triumphalist tenor of a capitalist modernization or socialist march into the future is if not completely replaced in the national atlases of the late twentieth century, at least colored by a more critical appreciation of the environmental costs of economic growth and the economic threat of environmental hazards and disasters. Comparing the 1989 *National Atlas of Jamaica*, for example, with the 1971 edition reveals a greater sensitivity toward environmental risks. There are maps of hurricane paths and of disaster-prone areas. Similarly, the 1978 edition of the *Atlas de Cuba* contains maps of the large hurricanes sweeping across the nation, a topic that does not appear in earlier editions.

At the end of the twentieth century, the growing environmental concern led to a growing number of national environmental atlases. One of the most sophisticated is the 2005 *Environmental Atlas of Sri Lanka* in which the national space is reimagined. There are maps of wildlife, land cover, wetlands and ecotourism (see Figure 6.14). There are maps of environmental concern that show maps of soil erosion, landslides and flooding. Cyclones, storms and

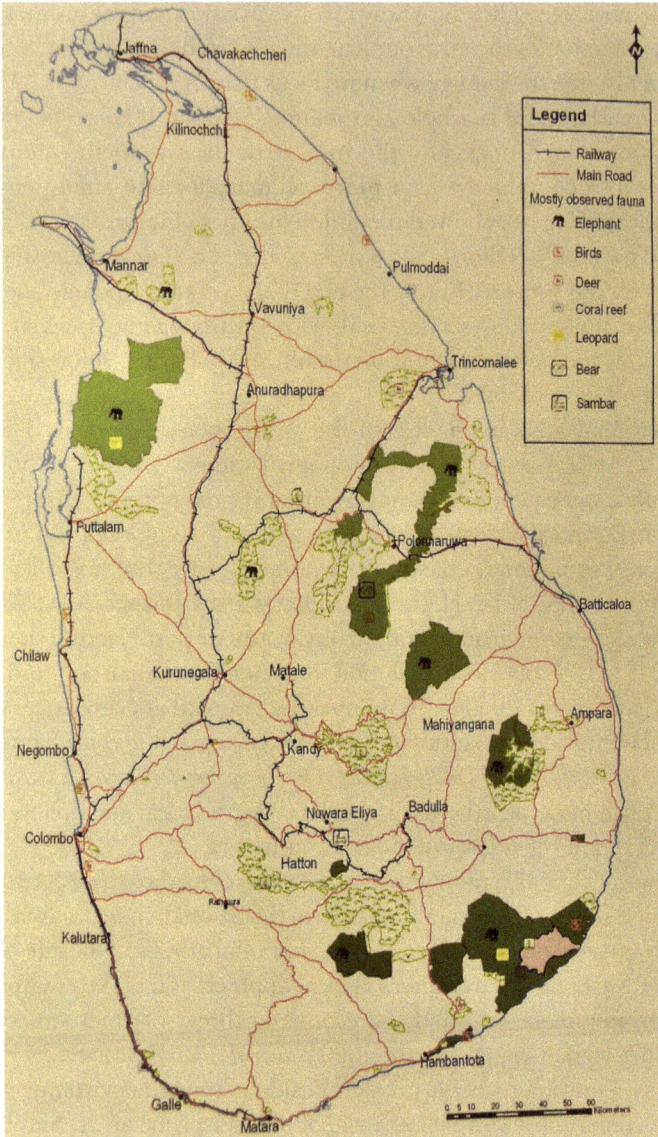

Figure 6.14 Ecotourism sites in Sri Lanka, *Environmental Atlas of Sri Lanka*, 2005. *Source:* Photo by John Rennie Short.

tsunamis are discussed. The atlas appeared after the 2004 tsunami, which wreaked havoc across 750 kilometers of coast. Rather than the traditional maps of agricultural products, there are maps of agro-ecological regions. Water and air quality are mapped, and the need for a national environmental education is stressed.

The physical space of the national territory is reimagined in the more recent national atlases. By the end of the twentieth century, many national atlases contained ideas of ecology, the interconnections between things, sustainability and the idea of a living earth, ideas that in many ways echo some of the fundamental ideas of the early cosmographers. The cosmographers' ideas of a living connected world imbued with a delicate grace, and Humboldt's notion of a life force both find expression in the concerns of late twentieth- and early twenty-first century national atlas and with their growing concern that our world needs to be managed and protected as well as measured and described.

Chapter 7

THE SOCIAL WORLD OF
THE NATIONAL ATLAS

The national atlas depicts the people of the nation-state. This depiction is not innocent of wider political considerations. The theorist Michel Foucault argued that human life is enmeshed in politics as a form of "biopolitics," a term he based on the work of Rudolf Kjellén, the Swedish political scientist who coined *geopolitics* as a term for the study of the intersection of human biology with politics. The state regulation of bodies in laws, rulings and practices is an integral part of the power of the state, whether it be in the right of entry to the country, the age of consent, the legality of abortions, the age of retirement with state benefits or the right to assisted suicide. The state lays claim over life and not just in the threat of death or incarceration. The state extends and deepens its reach over our bodies in myriad ways. Foucault wrote of a "power that exerts a positive influence on life, that endeavors to administer, optimize, and multiply it, subjecting it to precise controls and comprehensive regulations."[1] The modern state is conceptualized and materially produced through the management of the land and the people.[2] In the previous chapter we looked at the land, here we consider the people.

The national atlas was an important tool in biopolitics as it offered a visualization to the way that the population was enumerated, classified and ultimately controlled. To manage, control and regulate, it is necessary for the state to have a sense of how many bodies there are, where they are and what types. While the absolute amounts are a simple matter of arithmetic, the categorization is as much a social construction as a mathematical one. The national atlas gives us an insight into the realm of biopolitics because it employs a social categorization of the population that reveals much about the tensions and anxieties as well as the hopes and fears of the state, the political elites and the scientific establishment. The national atlas constitutes an important record of the constant and the changing classifications adopted, used and in some cases abandoned. As an example of the changing configurations, let us consider some of the population categories used in one of the earlier editions of the national atlas of Finland. As we will see, it is an interesting and somewhat surprising example.

If we compare countries in terms of human development, the countries of Scandinavia stand out. On the *Human Development Index* compiled by the United Nations, countries are rated on a composite index of quality of life, education and standard of living. In 2020, Norway ranked 1st, Sweden 7th, Denmark 10th and Finland ranked 11th in the world. The United States ranked 17th. Finland along with its Scandinavian neighbors is rightly considered one of the most socially progressive, rich and free societies in the world. It may come as a surprise then when looking at the 1925 *Atlas öfver Finland* to see maps with the title, "Insane and mentally defective." The maps use 1925 data to express this group per 10,000 inhabitants. The highest concentration, more than 15 per 10,000, is found in the southwest coastal areas between the urban areas of Turku and Helsinki (Figure 7.1). This edition of the national atlas is obsessed with deviancy and social disorder, with many maps of crimes against persons and crimes against property. It even has a map of the population of "deaf mutes." Another plots the deaths from pulmonary tuberculosis that is not included in the public health section, so it is used more as a measure of poverty/deviancy than health status. These types of maps disappear from subsequent editions of the national atlas. And, of course, the words "mentally defective" are never used again. That a bastion

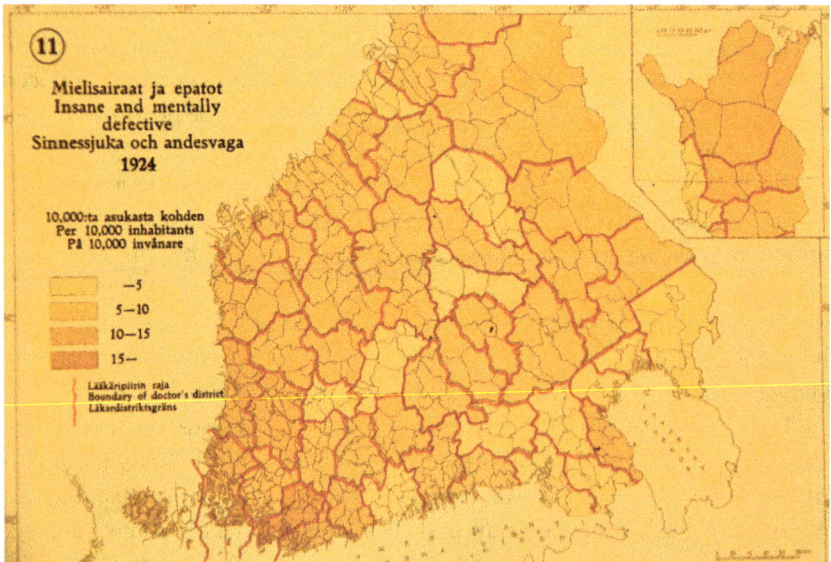

Figure 7.1 "Insane and mentally defective" *Atlas of Finland*, 1925. *Source:* Photo by John Rennie Short.

of liberal democracy and socially progressive policies should have such titles and topics in their national atlas is revealing; it highlights, in an extreme form, the notion of biopolitics. The national atlas of Finland was produced in 1925 at a time when eugenics was a powerful idea in Europe and the United States. The underlying assumption was that social behaviors were biologically determined, genetically coded in other words. The explanatory emphasis was on nature not nurture, so crime and deviancy were seen to arise from an unchanging genetic determination. It was in the blood. The goal was therefore to identify deviants and if possible, stop them from breeding. It was thought possible to improve the human population by selective breeding. It was necessary to identify "undesirables" and "social deviants" and restrict their ability to enter a country or if already here, stop them from having children. At about the same time as the atlas was produced in Finland, in the United States the 1924 Immigration Act was designed to keep out Italians and Eastern Europeans because they were considered genetically inferior. Their continued immigration would genetically weaken the country. In 1927, the Supreme Court of the United States upheld the right of states to forcibly sterilize persons considered unfit to procreate. In one of the court's worst decisions, *Buck v. Bell* 1927, Justice Oliver Wendell Holmes argued.

> to prevent our being swamped with incompetence. [...] It is better for all the world, if instead of waiting to execute degenerate offspring for crime, or to let them starve for their imbecility, society can prevent those who are manifestly unfit from continuing their kind. [...] Three generations of imbeciles are enough.[3]

The rise of the eugenics movement and the development of statistical techniques were closely bound together in the last third of the nineteenth century and the first part of the twentieth century. English statisticians such as Francis Galton (1822–1911) pioneered a more robust statistics with the development of correlation in 1888 that allowed comparisons across data sets. He also originated methods of statistical comparison such as "mean standard deviation" and "correlation regression." He applied most of these techniques to human characteristics. Part of his goal was to identify both norms and deviants. His protégé Karl Pearson (1857–1936) also worked on biometrics as well as on the development of chi-squared tests, histograms and correlation coefficients. Later, the psychologist Charles Spearman (1863–1945) developed the technique of *rank correlation*, which measures the rank of two sets of variables, let us say the height and weight of persons ranked from tallest to shortest and from heaviest to lightest. Pearson also developed factor analysis, which allowed researchers to collapse large data matrices into smaller factors without losing

too much information. I have used the technique myself in a number of different studies.[4] Both Galton and Pearson worked on data of human characteristics and were very influential figures in the early eugenics movement.

The 1925 edition of the *Atlas öfver Finland* was produced as ideas of eugenics,
identifying deviancy and controlling it, were gaining support in the United
States and across Europe. The classification of criminal activity and diseases
of the poor was part of an attempt to provide the data for this eugenics ideology. The Finnish population was reimagined in the 1925 atlas as a binary
of normal and deviant populations. As eugenics waned and was discredited
by science and by its overly eager adoption by Nazi Germany, in subsequent
national atlases, the Finnish population was no longer imagined such a way.
This Finnish example reveals how ideologies shaped how populations were
imagined, categorized and represented in the national atlas.

Social Statistics

Counting and especially "counting heads" are important parts of statecraft.
Numbers are needed to raise revenue, provide armies and allocate resources.
Such statistics even play a role in the foundational story of Christianity.
Joseph and Mary had to travel from Nazareth, where they lived, to Joseph's
hometown of Bethlehem because the Romans ordered the head count for
taxation purposes to take place in people's place of birth. Jesus was born in
Bethlehem, not Nazareth, because of the Roman census.

The population and other social maps of the national atlas drew on two
related discourses in the emergence of what we would now call *social sciences*.
The first is what we can call *social statistics*. An early influence is William
Petty (1623–1687). He was a key figure in the English colonization of Ireland.
One of his jobs was to survey the land confiscated by the English Crown and
then given to English soldiers. He developed the idea of *political arithmetic*.
Petty measured and made estimates of economic activity, including income
and wealth, labor productivity and population. On his estates in Ireland, he
surveyed the land, the labor and the livestock. He used the term *labor stock* for
the value of the people's work, sometimes interpreted as an earlier example
of Marx's *labor theory of value*. He estimated and averaged data, so he can also
be seen as an early statistician as well as a data gatherer. He was a forerunner
of economists such as Adam Smith and Karl Marx (at least the Marx of the
labor theory of value) as well as the social statisticians of the nineteenth century,
and his work is an early example of an empirical, measurable view of society linked to an instrumentalist view of government in an emerging capitalist economy. William Petty heralds the rise of the data-driven state. Petty's
empirical approach was linked to wider goals, including the creation of an

industrious and loyal population, a more efficient economy and—although wary of too much government intervention in private markets—he also saw the value of government spending. He predates John Maynard Keynes's twentieth-century argument regarding the need for the government to stimulate what we now call *aggregate demand* during economic downturns. Our reliance on measuring social and economic data is now such an accepted part of our world that it is difficult to see anything unusual about his writings. All government institutions and private organizations now use data to improve their performance: it is a practice defining the modern state so completely that it is difficult to grasp that it had to be invented and refined. William Petty was an important early figure in the creation of the social statistics that we now take for granted. His methods helped to create a new instrument for the state.[5]

A large part of the state was constituted in the generation and analysis of social data. By the twentieth century, the generation, compilation and analysis of social statistics were an integral part of government. Censuses and surveys of demographics, social characteristics and economic activity were embodied and embedded in a range of government agencies. Indeed, some agencies such as the U.S. Census Bureau were created with the express purpose of generating, collecting and analyzing social statistics. By the twentieth century, social data collection and analysis became a taken-for-granted role of government, forming the building material of the social worlds represented in the national atlas.

The display of social statistics

The second discourse of social science that shaped the national atlas was the display of social statistics and in particular social mapping. By the nineteenth century, Petty's political arithmetic was a standard government practice in the rapidly growing capitalist economies. Data was collected across a broad range of demographic, economic and social issues. To the long-standing data on births and deaths was added a range of new information on economic activity, public health and social behaviors. The interest in social statistics was fueled by concerns of fluctuating economies, growing cities, expanding populations and growing health issues. Rapidly growing cities of the nineteenth century helped forge ideas of public health as an organization and as a practice.

The rising levels of empirical data often overwhelmed the ability to analyze. The statistical tools to analyze increasingly large data sets were still in their infancy. In 1662, John Graunt (1620–1674), sometime known as the Columbus of biostatistics, used data from death records from London to make estimates about the city's total population and the spread of disease. He produced tables that gave the probabilities of survival. However, statistical modeling beyond the simply descriptive had to wait until the late nineteenth century, when new statistical techniques allowed data to be compared in a more rigorous fashion. But before the full development of these robust statistics, data display and social mapping were the most important ways to suggest causal connections. William Playfair (1759–1823) is an exotic figure whose resumé includes the categories of engineer, political economist, land speculator, convict, blackmailer, failed banker and secret agent.[6] The Scottish engineer managed to be both a secret agent for the UK in their war with France as well as an original creator of the graphical display of statistics. He introduced numerous methods to display social data, including the pie chart, the line chart, the bar chart and the histogram. His *Commercial and Political Atlas* of 1786 (the third edition was printed in 1801) contains numerous graphic representations of economic activity. While fellow Scotsman Adam Smith was providing a theoretical justification of the capitalist market, Playfair was detailing its spatial reach in graphical representations. Despite the title, his *Atlas* does not contain any maps. Instead, it is filled with diagrams displaying British trade with other nations. One graph shows the exports and imports to and from the West Indies: it is a figurative display of an emerging global capitalism centered on Britain.[7] The full emergence of a capitalist economy was prompting graphical consideration. Maps of global capital flows appear in many subsequent national atlases, including the *Bol'shoĭ sovetskiĭ atlas mira* (*Great Soviet World Atlas*) of 1937 (see Figure 4.3) and in the 2000 *Atlas Nacional do Brasil* (see Figure 5.3).

Playfair provided ways to display and hence understand economic data. His diagrams continue to be used today. He did not employ maps but in the nineteenth century maps were increasingly used to display statistical data. *Thematic mapping*, the use of maps to display specific themes such as crime, mortality or disease began to advance. The Frenchman Charles Dupin (1784–1873) produced perhaps the first choropleth map. These are maps that display different sets of values of the same variable across different territorial subdivisions. In 1826, Dupin used this technique to map illiteracy levels across the departments of France. His shading went from light to dark in proportion to the rising degree of illiteracy (Figure 7.2). It gave the map a symbolic feature. It was often referred to as the "Map of Dark and Enlightened France." The map and its author achieved widespread recognition for the novel presentation of statistical information: it is mentioned by Balzac and Stendhal and achieved widespread recognition outside France.

Moral statistics

The term *moral statistics* first appeared in an essay written in 1833 by the French lawyer and part-time statistician André-Michel Guerry (1802–1866). The term refers to levels of crime, deviancy and poverty. "Crime" maps first

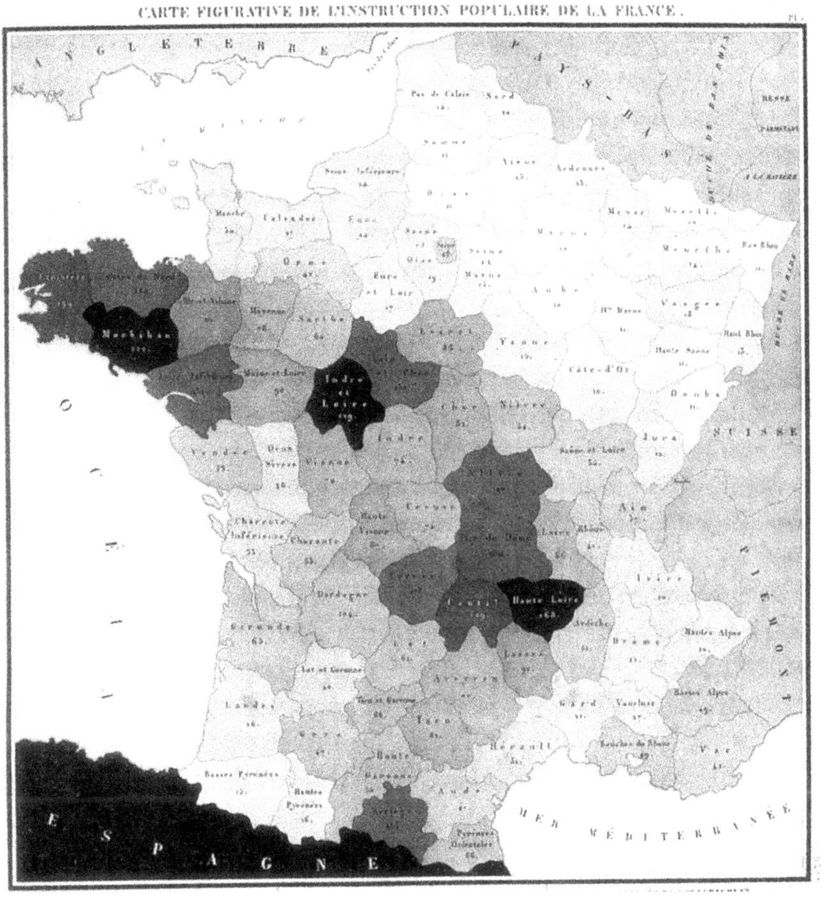

Figure 7.2 Map of France, 1826. *Source:* https://en.wikipedia.org/wiki/Charles _Dupin#/media/File:Carte_figurative_de_l'instruction_populaire_de_la_France.jpg.

appeared in France in 1829 when Guerry, along with the Venetian geographer and statistician Adriano Balbi (1782–1848), used data from 1825 to 1827 to plot, for each of the départements in the country, the incidence of crime against persons and against property in relation to "educational instruction." Their maps sought to demonstrate the correlation of crime and educational levels. The Belgian Adolphe Quetelet (1796–1874) introduced the notion of the "average man" and used the idea of deviance from this norm and produced maps of the crime rates, marriage rates and suicide rates. He also, by the way, developed the "body mass index," which was originally referred to as the Quetelet Index. Quetelet also struck up a relationship with the prominent nurse Florence Nightingale, and they shared a similar view that displaying social data could lead to social progress and improvement. She produced an interesting and innovative chart depicting the causes of deaths of British soldiers during the Crimean War. Nightingale's diagram, described as a "beautiful and persuasive call to action," clearly reveals that most deaths resulted from infections rather than wounds.[8] The diagram ultimately led to improvements in military hospital care.

In 1864, Guerry went on to produce sophisticated maps of specific types of crime such as murder, rape, "theft by servants" and suicide. His work was read and admired in Europe and the United States. He laid the basis for later work such as the classic work on suicide by the Frenchman Emile Durkheim whose 1897 book, *Le Suicide*, is widely accepted as a sociological classic. Durkheim discussed the topic as a social issue and not one of personal anguish. He hypothesized that Protestants would have a higher rate of suicide than Catholics because Protestant society had lower levels of integration. He tested his assumption with aggregate data. Durkheim's study is widely acknowledged as a foundational text of the discipline of sociology.

Social mapping was used to make inferences about the causes of phenomena from their codistribution with other data. Guerry, for example, was convinced that criminality was caused by the lack of education but changed his opinion after comparing maps of the two variables, which showed little correlation. We should note, however, that his conclusions, and those of Durkheim, are undercut by "the ecological fallacy," which is the mistaken belief that aggregate-level data, such as suicide or crime rates across countries and regions, may be used to explain individual attributes. Firm conclusions can only be made with individual case study data and not aggregate data. But, before more sophisticated statistical techniques were developed, and more individual data were available, social mapping was able to give hints and suggestions about causal connections. Before the onset of this statistical modeling, social mapping allowed the first simple testing of hypotheses and ideas about social order and disorder.

Maps, along with other forms of visual displays such as charts and figures, summarized and made more comprehensible vast amounts of accumulating social data. Compared to columns of tabular data, visual forms allowed easier comparison and the possibility of identifying causes as well as effects and provided a basis for not only measuring society but also explaining it. Social mapping was an important part of the visual expansion of social analysis.

The rapid growth of towns and cities in the nineteenth century was often a cause for alarm among commentators. While Karl Marx and Friedrich Engels saw the rapid growth of industrial cities as a source of revolutionary change and a cause for optimism, many others were worried by the prospect.

The city became a popular site for moral statistics and social mapping. As cities increased in size, the rich and the poor tended to be separated into different neighborhoods, and "urban segregation" became a subject of cartographic interest. For example, the 1841 census of Ireland provided the basis for a whole series of maps concerning social characteristics, including a map of Dublin that color coded the streets as "high class," "first class," "second class" and "third class"—the latter being deemed to be inhabited by "artisans, huxters and low population." A similar type of classification was adopted by Charles Booth (1840–1916), a rich shipowner and social commentator, in his 1889 study of London entitled *Life and Labour of the People in London*. This 17-volume work contains several maps of social classification. The "worst" areas are said to be inhabited by the "very poor," and Booth further characterized these people as "lowest class" and "vicious, semi-criminals." Just like Dupin, his categories went from light to dark as one went down the social hierarchy to the criminal underworld. The maps provide us with an interesting picture of urban life in the nineteenth century because the classifications used and their underlying ideologies tell us much about the worldview of middle- and upper-class commentators of the day. For many of them, cities were "social volcanoes" that could erupt at any time, and urban mapping was an attempt to predict where the "social seismic activity" was strongest and the threat of anarchy the most extreme.

Maps were used as a way of plotting, and hence understanding, social change and its consequences, especially in the cities[9] where rapid industrialization and urbanization led to major centers of disease and transmission. The authorities began to collect and to centralize medical data for analysis. A particular concern was public health, and in Britain the Poor Law Commission published its *Report on the Sanitary Condition of the Labouring Population* in 1842. The report included two maps of housing types and incidence of disease in Leeds and in the Bethnal Green neighborhood of London.

Maps were used to "plot" diseases and epidemics. A worldwide cholera epidemic began in India in 1817, and the American mapmaker Henry Schenck Tanner (1786–1858) produced *A Geographical and Statistical Account of the Epidemic Cholera from its Commencement in India to its Entrance into the United States* in 1832. Tanner's book consists of a series of tables listing the number of cases reported, duration in days of the "pestilence" and the number of deaths in different localities, by country. A world map then shows the diffusion of the epidemic from India in 1817. Tanner also produced a more detailed map of the United States and New York that shows the sites where cholera had broken out by means of small red dots. This map has a detailed chart of the diffusion of the illness, showing how the disease spread up the Hudson River and along the Champlain and Erie Canals.

One of the most famous medical maps was drawn by John Snow (1813–1858), a doctor working in mid-nineteenth-century London.[10] He was convinced that cholera was communicated by contaminated water, and in 1855 he published *On the Mode of Communication of Cholera*, which contains two maps. The first shows two areas of London served by two different water companies that used different sources for their water supply. While the area served by one company had death rates due to cholera of only 5 per 1,000, the other had rates of 71 per 1,000. Snow's second map plotted the distribution of cholera cases and showed that they clustered around particular pumps—for example, people using the water pump in Broad Street were more likely to go down with cholera. It was a convincing—and correct—argument.

The French engineer Charles Joseph Minard (1781–1870) was a key figure in thematic mapping.[11] He developed what he termed his *carte figurative* after his official retirement at the age of 70 in 1851. He created more than sixty statistical graphics that captured the economic and social changes of the industrial revolution in a globalizing capitalism. Minard made flow maps, for example, of railway passenger traffic and French trade links. Figure 7.3 shows road traffic between Dijon and Mulhouse. (Compare it with other flow maps in later national atlases as depicted in Figures 4.5 and 5.9.) His main subject matter was the economic geography of capitalist economies. He pioneered a cartographic political economy that mapped the flow of imports and exports, which later became a staple cartogram of the modern national atlas. He produced maps with graduate circles, pie charts and bar graphs. Minard gave cartographic form to the statistical display tools developed earlier by Playfair. He was unconstrained by the rigid grid of latitude and longitude. The nature of the data being represented sometimes forced Minard to change from absolute to relative space. To fit in the thick flow lines representing commerce through the narrow English Channel, for example, he widened the channel out of proportion to the land surrounding land masses. Minard felt free to create relative spaces rather than being restricted by absolute space.

Minard added a spatial dimension to time series data. His most famous *carte figurative* is his 1861 map of Napoleon's advance and retreat from Moscow in 1812–13.[12] The graphic plots six variables, including the changing size of the French army, its movement and the temperatures along the route during the brutal winter campaign, The dean of data visual display, Edward Tufte, describes it as "the best statistical graphic ever drawn."[13] Maps in the style of Minard became an important element in the cartographic practices of the national atlases of the late nineteenth and twentieth century. The 1970 *National Atlas of the United States*, for example, has maps of imports and exports that draw inspiration directly from Minard.

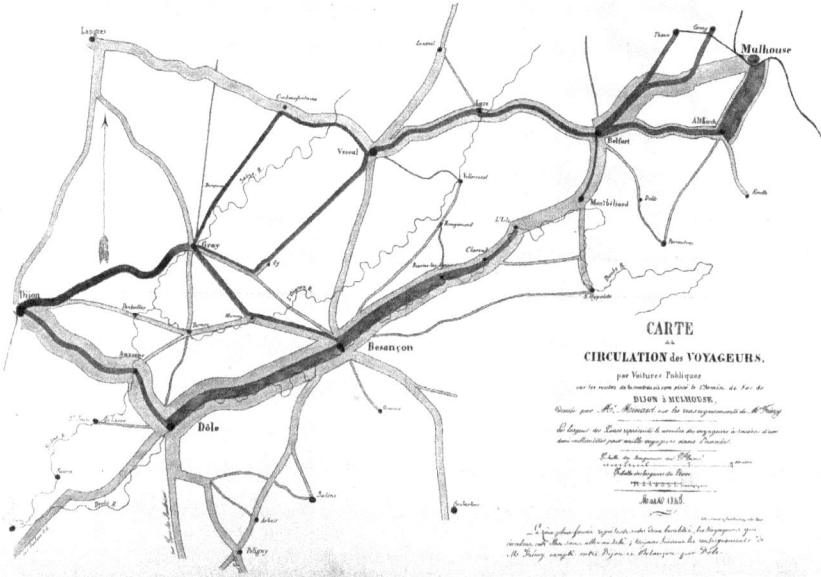

Figure 7.3 Road traffic in France, 1845. *Source:* https://commons.wikimedia.org/wiki/File:Carte_de_la_circulation_des_voyageurs_par_voitures_publiques_sur_les_routes_de_la_contrée_où_sera_placé_le_chemin_de_fer_de_Dijon_à_Mulhouse,_1845.jpg.

The techniques of statistical display and social mapping spread through international collaborations, exchanges and formal, scientific meetings.[14] Dupin's choropleth map design became used widely and adopted quickly in Britain. His technique was used in the mapping of disease outbreaks in London as well as in Paris. Charles Booth used a similar shading design in his maps of social class in London. Both Dupin and Guerry were made members of the Statistical Society of London. Thematic mapping was standardized across different nations in international scientific meetings. The newly formed International Institute of Statistics held meetings between 1869 and 1901 in Budapest, St. Petersburg and The Hague. In 1914, the American Statistical Association issued a set of formal recommendations about statistical display techniques.

It was a time of social change and turbulence, including the growing marketization of economic activities, the vast rural to urban migration, growing health concerns and rising social inequalities in rapidly growing cities. Social statistics were deployed to describe and to explain these new threats to social order. At the same time, new printing technologies allowed a range of print media including books, newspapers and magazines to show the visual display of quantitative data in figures and maps. Maps and diagrams became part of the print capitalism of the nineteenth century and entered the realm of public discourse of an increasingly educated population.

Statistical Atlases

The growing fascination with statistics as the nineteenth century progressed—especially with social statistics—meant that at the intersection of statistics and the national atlas was born the national statistical atlas. The heyday of this variant was from 1875 to 1915, when national statistical atlases were printed for Argentina 1873, Russia 1873, United States 1874, German Empire 1876, France 1878, Portugal 1881, UK 1882, Austrian Empire 1882–87, Mexico 1886, Belgium 1899, Japan 1902 and Prussia 1905.[15]

Statistical atlases of the United States

We can consider one national example, the statistical atlases of the United States. Six in total were printed, in 1874, 1883, 1898, 1903, 1914 and 1925. The first *Statistical Atlas of the United States* was printed in 1874 and based on data from the 1870 census.[16] The subsequent statistical atlases were based on data from the 1880, 1890, 1900, 1910 and 1920 censuses, respectively. To the basic headcount of the early censuses was added a growing inventory of social, demographic and economic information.

The moving force behind the first U.S. statistical atlas of 1874 was Francis Walker (1840–1897). He was a Boston Brahmin who held important political and academic posts and wrote and lectured on a range of issues of contemporary concern. He held a professorship at Yale before becoming president of Massachusetts Institute of Technology from 1881 to 1897. He was actively involved in professional associations as vice president of the National Academy of Sciences, president of the American Economic Association and president of the American Statistical Association.

He had a keen interest in social statistics, especially in moral statistics. He was appointed Chief of the Bureau of Statistics and made the Superintendent of the 1870 census. He lasted just over a year in this position as his political patronage evaporated. He was then made Commissioner of Indian Affairs from 1871 to 1873. In response to the problems of the remaining Indian tribes blocking westward expansion, he proposed consolidating them in large reservations, where they would be safe from white incursions, with facilities for training and education. It was profoundly racist but more enlightened than many other proposals at the time, including some that called for outright annihilation.

Soon after arriving at Yale in 1873, he argued for a much more sophisticated illustration of the census data than had just been released. He and colleagues at Yale managed to persuade the Secretary of Interior to write to Congress about the need for and importance of graphical and cartographic illustration of these data. Congress agreed and appropriated $30,000 for the atlas in March 1873. Walker took leave from Yale to complete the project. It was published in 1874, and 5,000 copies of the atlas were sent to public libraries, colleges and learned societies to inform public opinion and advance political education.

This atlas is the ultimate in East Coast bias with many maps only showing the country east of the 100th Meridian. The atlas is especially strong on social data, including population distribution and density, health, immigration, religion, ethnicity and economic trends. It contains 60 plates of maps and diagrams that are discussed in some detail by Walker. In the introduction, he outlines how to read them, drawing attention to understanding the intensity of shading, the distinction between absolute and relative proportions and the need to see the diagrams as a whole to make connections. Part 1 of the atlas was devoted to physical features, and Part 2 to the population with maps of the black population, foreign-born, illiteracy and religious affiliation. He produced maps of both absolute and relative proportions of different national origins. The resultant maps allow us to see how the Irish are concentrated in the North and Northeast, and the German presence is particularly pronounced in the upper Midwest. Part 3 details the moral statistics with maps and diagrams depicting mortality and afflictions, including blindness, deafness and insanity. There is a focus on social difference, abnormality and especially illness. Walker was writing at a time when social control and social surveillance were important discourses in a nation experiencing rapid urbanization and large-scale immigration.

The atlas is an incredibly sophisticated display of social statistics. Some of them are still in use. Walker identified the center of the national population. This has been depicted in all subsequent atlases, showing a steady westward movement with each successive census. Some of the statistics influenced wider debates. Walker uses the term *frontier line* to demarcate areas of low population density, defined as less than two persons per square mile. The term *frontier* became part of the language of the census and entered the national mythology of the United States. When the superintendent of the 1890 census declared that the frontier was over because of increasing population density, a young ambitious historian Frederick Jackson Turner took the declaration as the basis for a thesis he presented at a history conference in Chicago in 1893. The title of his paper was "The significance of the frontier in American history." He argued that a distinctly American democracy was forged along the frontier. This idea has generated debates ever since about the role of the frontier in shaping the history and more generally the identity and character of the United States.

The 1874 atlas was part of a broader attempt in the nineteenth century to provide a cartographic and statistical basis to better understand society. The atlas was not just an inventory but a way to search for connections and specifically to identify sources of disorder, especially crime, disease and their relationship with the "other." As superintendent of the 1870 census, Walker added tables on school attendance, illiteracy, pauperism and crime. Among the social categories he identified in the atlas were the blind, deaf, dumb, insane and idiotic. In one section of the atlas, he draws attention to the blind, deaf and mute, and "insane idiots," who were disaggregated by age, sex, race and nationality. Figure 7.4 shows a complex diagram that depicts "idiocy" broken down by state gender and race and nativity. Despite the seemingly scientific basis of the figure, it is firmly based on the pseudo-science ideological enterprise of eugenics.

As a typical upper-class social observer of the nineteenth century, Walker had an enduring concern with social order, control and surveillance of the "other." The atlas reveals not only the social geography but also the social and political concerns of the intellectual and economic elite of the time with a focus on social difference and a concern with abnormality. Underlying the atlas in all its cartographic and sophistical innovations was a desire for greater knowledge of society, a knowledge connected to social power. The sense of searching for the locations and causes of social disorder, very broadly defined, lies at the very heart of Walker's statistical rendering of the nation.

Despite the sophistication of the diagrams and maps, the data source for the atlas—the 1870 census—was one of the most inaccurate. It was undertaken by U.S. marshals, who were political appointees, and not skilled census workers. The populations of Indianapolis and Philadelphia, for example, were counted twice. The returns were unreliable in the South because the 1870 census was the first after the civil war, and there was a huge noncompliance by many whites protesting the exercise of a northern federal power. It is estimated that 1.2 million white southerners were missed by the census. The next census, the 1880 census, was more accurate as the U.S. Census Office had direct control over hiring skilled enumerators. The chief geographer of the 1880, 1890 and 1900 censuses was Henry Gannett (1846–1914), another New England savant who had membership in the American Statistical Society, American Economic Association and the National Geographic Society. Gannett was coauthor of the second statistical atlas, *Scribner's Statistical Atlas of the United States*. This atlas was based on data from the 1880 census and printed in 1883 by a private company. The subtitle reveals its intent, "Showing by graphic methods their present condition and their present political, social, and industrial development." The book was dedicated to Francis A. Walker, with the inscription, "to whom the country is chiefly indebted for

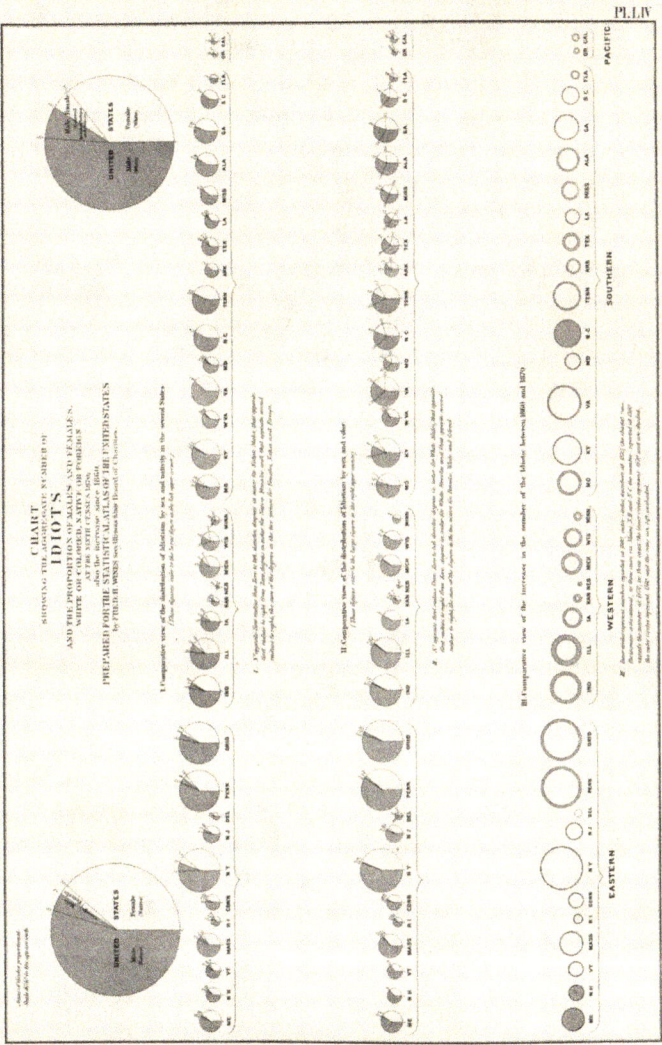

Figure 7.4 "Idiocy," *Statistical Atlas of the United States*, 1874. *Source:* https://www.loc.gov/resource/g3701gm.gct00008/?sp=107.

a thorough knowledge of its present condition and resources." The population section of the atlas used a complex categorization of race and nativity to identify colored, foreign and then various nationalities. One diagram has a simple breakdown of "Born in the United States," "Foreign-Born," "Colored," "Germans," "Irish," "English," "Canadians," "Swiss" and "French." There are diagrams and maps recording the nativities in the main cities. One series of maps has the title of "defective, dependent, and delinquent." Under this alliterative grouping are separate maps for the insane, idiotic, blind, deaf mutes, paupers and prisoners. There are also maps of diseases, illiteracy and religion. Gannet was also the main author of the 1898 and 1903 statistical atlases.

Statistical atlases of the "new" Baltic states

The heyday of the statistical national atlas was coming to an end by the beginning of World War I. Only one statistical atlas was published in the United States, after World War I, in 1925. However, in the immediate post-war period, there was a brief flourish in some of the new nations that emerged from the fall of the German and Russian Empires. In 1918, the Baltic countries of Estonia, Latvia and Lithuania obtained independence. They celebrated the subsequent anniversaries with a statistical national atlas in their respective national language of the countries, as well as in French. The atlas for Estonia, *Eesti Statistiline Album / Estonie Album statistique*, was published in three volumes from 1925 to 1928; the Lithuanian atlas, *Lietuva Skaitmenimis / La Lithuanie en chiffres* in 1918–28; and Latvia's *Latvijas Statistikas Atlass / Atlas statistique de la Lettonie* in 1938. They were all high-quality productions in full color using the latest chromolithographic methods. They draw on the mounting social statistical information collected by the nation-states. Produced to high standards, these large format products, mostly printed in full color, were prestige publications meant to provide a statistical-cartographic profile of the nation-state, awaken national consciousness, educate a national population and provide an informative text for foreign audiences.[17] They were expensive undertakings, prestige publications meant more for elite audiences in the respective countries and especially for overseas, rather than as everyday teaching texts in local schools. Produced to roughly coincide with the 10- and 20-year anniversaries of independence, they were also used to provide a portrait of themselves as new nations and as an integral part of a new, postimperial Europe. Tragically, the aspirational attempt to secure at least in textual form, a greater hold of their independence, was dashed when they ceased to exist in 1940 after they were folded up into the Soviet Union. These three atlases are still to this day the only national atlases of the three countries and remain as one commentator noted, "masterpieces of cartography and exemplary representations for the era of graphical statistical atlases."[18] Figure 7.5, for example, captures a wealth of demographic data in maps and a variety of statistical figures.

Social Categories

Populations are not only counted and mapped in the national atlas, but they are also classified and categorized. The classification systems used, and the categories employed, tell us much about changing views of society. The social categories are not objective, immutable, given facts; they are social constructions that tell us about racial insecurities, ethnic tensions and changing attitudes toward the "other." The census and atlas do not enumerate and display social categories; rather, they create them.

As we have seen, the eugenics movement cast an influence over the social data produced, employed and mapped in most national atlases printed between 1870 and 1950. For example, the various *Statistical Atlas of the United States* and the 1925 *Atlas öfver Finland* are filled with so-called deviant populations defined in relation to physical and mental health (see Figures 7.1 and 7.4). Eugenics waned after World War II and the emphasis of atlases shifted back to more basic demography data with the addition of health status and education as governments became more involved in the provision of public health and public education. However, there were some categories that remained a consistent theme of the modern national atlas. There was the basic data of how many people were living and dying across the country. There were also the social characteristics of the population. These included racial and ethnic identity, and while age and gender were relatively fixed categories, these categories were more mutable as attitudes changed. Let's look at some of the dominant forms of classifications in the national atlas and how they changed over time.

Race and ethnic identity

Race is more of a social category than a biological fact. It is socially constructed. That does not make its effects any less real, but simply to draw attention to its basis in society more than in biology.

Race plays an especially important role in the national atlases of colonial and settler societies. In Chapter 3 we looked at the cartographic anxieties of newly independent countries in Central and South America. Let us revisit the early national atlases of Venezuela, Mexico and Peru. A table in the 1840 *Atlas Físico y Político de La República de Venezuela* lists six different groups. Translated

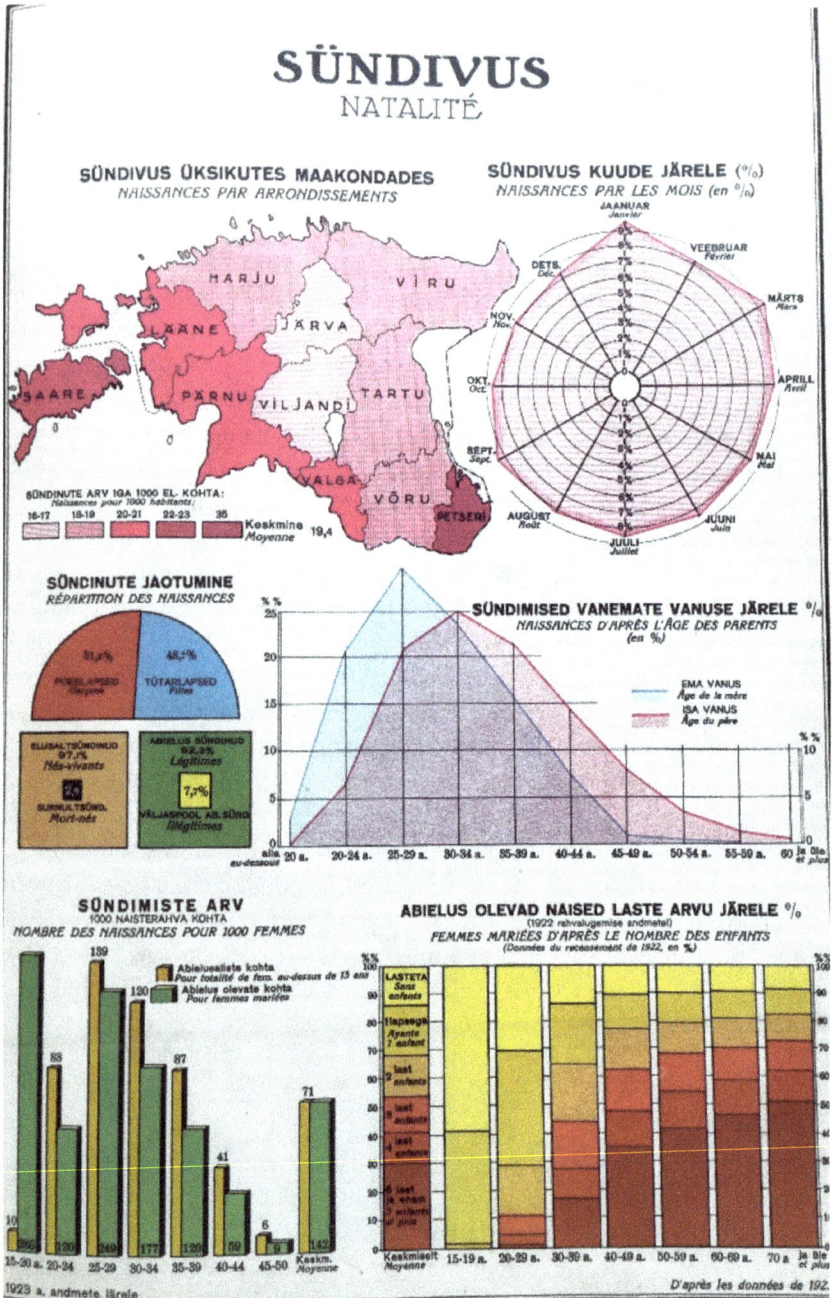

Figure 7.5 Births in *Eesti Statistiline Album / Estonie Album statistique*, 1925–28. *Source:* Photo by John Rennie Short.

into English they are white, mixed breeds, slaves, civilized Indians, baptized Indians and independent Indians. It is a mixture of race, status and religious affiliation. The categories used in the table allude to the power of the whites over the slaves, the liminal category of mixed races and an indigenous population fracturing along a path of assimilation. The term "civilized" as a category of Indian, rather a mark of progress more accurately represents the cowed and the defeated, and along with the category "baptized," represents those brought under control and no longer a threat to white Hispanic power.

Whites are at the top of the hierarchy. Indeed, the hierarchy is constructed around this apex with the lower the level, the more distant in status, wealth and ultimately power from this powerful elite. The term "mixed breeds" is a translation from the original Spanish of "Razas mistas." The colonial appropriation of the New World also involved a sexual appropriation. The bulk of immigrants from Europe in the early years were men. Sexual liaisons, some consensual, many forced, between Europeans and the local indigenous resulted in a category often referred to as mixed race or mestizos. Slaves were simply "slaves," and the underlying assumption was that they were black and originally from Africa. The indigenous population was further divided into "civilized and baptized" and the more independent "indigenous." This last group was the holdouts resisting incorporation or induction into the Christian faith. One map in the atlas depicts in rich detail the different tribes classified by language. The map shows a densely settled, richly diversified indigenous occupation across the territory of the new nation-state.

The depiction of indigenous tribes is also apparent in the 1885 *Atlas pintoresco é histórico de los Estados Unidos Mexicanos* with its ethnographic map and images of the different groups (see Figure 3.3). The categorization includes white race, indigenous race and mixed race (raza mexclada). The 1865 *Atlas Geográfico del Perú* depicts a land populated by different indigenous groups including different tribes such as the *Indios Cocamas* and *Indios Yameos*.

The categories used are both fixed and liminal. Fixity is the feature at the very top and very bottom of the racial categorizations. Whites are the top of the hierarchy with slaves at the bottom; slaves are neither given race nor language, with their description only referring to their power or subordination. Indigenous groups are at least given some recognition of their diversity. Mixed races are the liminal group, an open-ended categorization that attests to the sexual nature of colonization.

Race was a particular concern of the settler-colonial and slave-based societies and found a central position into many national atlases. The recurring population characteristic that remains central in all the national atlases of the United States is race and especially the racial division between black and white. It figures as a central classification in the six statistical atlases of the United States produced from 1874 to 1925. Figure 7.6, for example, plots the distribution of the "colored population" from data supplied in the 1870 census. Notice the heavy concentration of blacks in the South, their concentration a historical legacy of the location of slave-based agricultural economies along the banks of the Mississippi and the coastal and piedmont areas of southeastern United States.

We get some idea of the ideological basis by reconsidering the work of Henry Gannett. We have already come across him. He was founding member of the National Geographic Society and was chief geographer of the U.S. censuses of 1880, 1890 and 1900 and involved in the statistical atlases of 1883, 1898 and 1903. He also wrote *Statistics of the Negroes in the United States* published in 1894. In the *Introduction* he notes that there were 61 million whites and 8 million Negroes. The Negroes, he writes, "were brought into close association with a more civilized and stronger people." It is to be noted that there is no mention of slavery—only that the contact with whites would be beneficial to the "negroes." A few pages later he notes that the birth rate is lower for negroes than whites and "should set at rest forever all fears regarding any possible conflict between the two races." He repeats the old trope that "the proportion of criminals among the negroes is much greater than among the whites." The rate of incarceration was four times higher for blacks, but that was more due to a racist system than a predisposition to criminality. In one revealing remark, however, he also noted that "the commitments of negroes are for petty offenses in much greater proportion than among the whites." This, I think, is indicative of the white criminal system to punish blacks more than whites for small infractions. Gannet's casual racism is indicative of the context in which the statistical atlases of the United States were produced and for revealing prevailing attitudes about race and blacks.[19]

By the 1970 *National Atlas of the United States*, the toxic obsession with race had somewhat declined. In part because the atlas was under the authority of the Geological Survey which is concerned with environmental issues more than social issues, However, even with his physical science emphasis, there are still maps of the "negro" population including a map of change in the "negro" population from 1940 to 1960 (Figure 7.7). The map shows the drift away from the Old South to the North and West.

Race and ethnicity were central elements in the social categories represented in the atlases produced in colonial context. The *Atlas of Sierra Leone*, produced in 1953 while the country was still a colony of the UK, has maps that show the various tribal groups, including the Mendi, Kono and Kissi. The map also contains a detailed hierarchy of chiefdoms. The British needed to identify tribes and tribal leaders to forge relations with collaborative elites and ensure colonial control through the mechanism of indirect rule. The first

Figure 7.6 Colored population, *Statistical Atlas of the United States*, 1874. *Source:* https://www.loc.gov/resource/g3701gm.gct00297/?sp=66&r=-0.155,0.705,1.086,0 .543,0.

and second editions of *Atlas of Kenya*, printed in 1959 and 1962 respectively and produced under British colonial control, have land divisions that reference ethnicity and race. It distinguishes between Crown land, Native Settlement areas, Native Reserves and Native Leasehold Areas. It is a complex categorization of landownership and race. After independence, in the first atlas of an independent country, the 1970 *National Atlas of Kenya*, a new threefold land division was employed: "Government Land," "Trust Land" and "Private Land." The Trust Land was known as Native Reserves in the previous colonial era atlases.

The 1960 *Atlas of the Union of South Africa* was produced during the height of the apartheid era in South Africa that lasted from 1948 to 1994. Apartheid, essentially Dutch for "separate," was a racialized system of control, which identified three main racial groups whites, blacks and colored, and strongly enforced segregation between them. Blacks could work, but not live in the cities and had to leave before sunset. They were forced to live in townships on the edge of the city with very little in the way of public provision of physical and social infrastructures. Whites, both English and Dutch speakers (Afrikaners), lived more privileged lives while blacks were exploited for their labor and denied effective citizenship. The 1960 atlas was published in both Dutch and English, although the opening pages are all in Dutch, a reflection of Afrikaner political dominance. The atlas displays four major

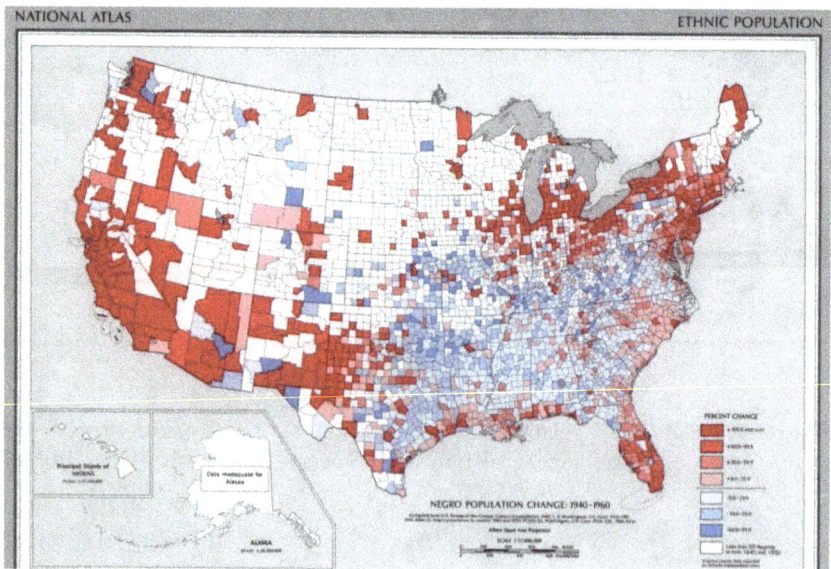

Figure 7.7 Negro population change. *National Atlas of the United States*, 1970. *Source:* https://www.loc.gov/resource/g3701gm.gct00013/?sp=191&st=single.

groups identified in the 1951 census: Europeans or whites, Natives, Asiatics and Coloreds. In some maps the European category was further divided into English- and Afrikaners-speaking. An even finer division identified the following language groups: Afrikaans only, English only, bilingual and neither. There was also a finer division of the Native (i.e., black) category into distinct tribal groups such as the Bantu, Xosa and Zulu. The racial obsession to distinguish between whites and others is shown in this definition of whites employed in the atlas as "persons who in appearance obviously are generally accepted as white persons but excluding persons who although in appearance obviously white are generally accepted as colored persons."[20]

The social categories of national atlases sometimes often combine categories of race, ethnicity and language with degrees of separation from the central authorities. The 1997 *Atlas of Pakistan*, for example, distinguishes between distinct language groups (see Figure 5.4). This is not a simple exercise in linguistic geography as the categories employed overlap with differing degrees of loyalty to the central government. The Pashtun-speaking areas, referred to as Pushto in the atlas, and the Baluchi areas have complex relations with the dominant Urdu-speaking political elite. The Pashtun-speaking area has a long history of sturdy independence while a long-running separatist movement in Baluchistan occasionally erupts to threaten Pakistan national coherence.

Race, ethnicity and religion are fluid categories that can also be revealing of underlying tensions in a society. The changing and elastic nature of racial categorizations is evident in the different editions of numerous national atlases. In the early censuses of Sri Lanka, for example, the population was divided along divisions of race, nationality and caste. But, after 1963, the term "ethnic groups" was used. Figure 7.8 from the 1988 *National Atlas of Sri Lanka* is a map of ethnic groups. It is an interesting and revealing categorization. It identifies Sinhalese (mainly Buddhist) and Tamil (mainly Hindu). It also differentiates between "Sri Lankan Tamil," those with a long ancestry on the island, and "Indian Tamil" to include immigrant workers from the Tamil-speaking areas of southern India who came from 1830s onward to work on the coffee and tea plantations in the highland areas. The category "Moor" is an interesting one, with Shakespearian connotations. It is a bilingual minority most of whom can speak both Tamil and Sinhalese and is predominantly Muslim. Some trace their ancestry to Arab traders who sailed with the trade winds across the Indian Ocean to first settle a thousand years ago, although Tamil nationalists sometimes dispute the Arab origin story seeing most of the Moors as Tamil converts to Islam and the Moor category as a Sinhalese effort to separate ethnic Tamils. Sri Lankan Tamils are mainly located in the north and northeast of the country. A Tamil liberation movement, popularly

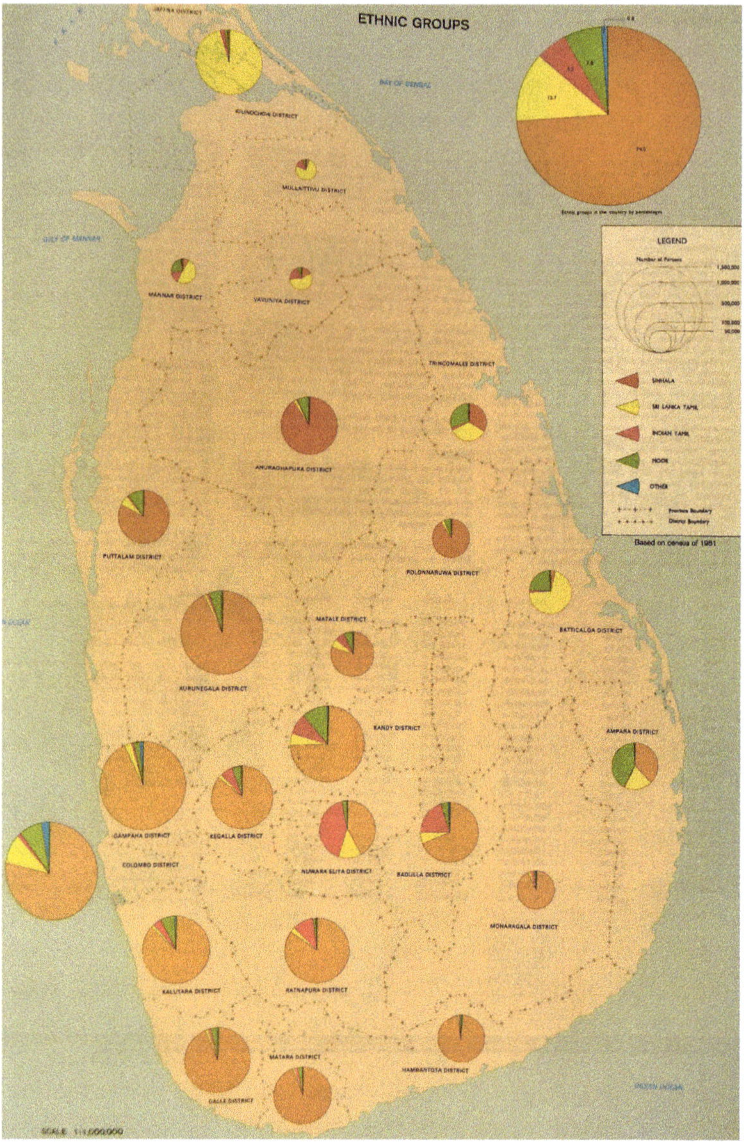

Figure 7.8 Ethnic groups, *National Atlas of Sri Lanka*, 1988. *Source:* Photo by John Rennie Short.

known as the Tamil Tigers, fought an insurgency campaign from 1983 to 2009. More than 100,000 died in the struggle and there are still more than 100,000 refugees in the southern Indian state of Tamil Nadu. During the conflict there were reports of the ethnic cleansing of Sri Lankan Moors by the Tamil Tigers in the northern provinces. The atlas was conceived and produced at the height of the political instability. Even since the peace agree-ment of 2009, conflict between the majority Sinhalese and the minority Sri Lankan Tamils still constitutes a major stress fracture in Sri Lanka's return to political stability.

The other

The "other" is not a general category employed directly in national atlases, but often implied. By "other" I mean a group considered different from the ruling majority. They are often pictured not just as different but often as a threat. The sociologist Stanley Cohen identified what he termed folk devils and moral panics. Folk devils are "the other" as threat. Moral panics are movements that are a response to putative threats to society. Moral panics tend to be virulent during periods of abrupt social change and economic dislocations.[21]

There are many examples of the *other* in national atlases. The 2006 *al-Aṭlas al-waṭanī al-Qaṭarī* (*Qatar National Atlas*) hints at the division between citizen and migrant with a map of labor camps for foreign workers. In the colo-nial era atlas of Senegal produced in 1925, *Atlas des Cercles de L'A.O.F*, the social categorization of "Races" only included distinct tribal groups such as the Ouolofs and Maurees. The 1977 postcolonial *Atlas National du Sénégal* employed a more complex categorization of ethnic, religious and linguistic groups, including traditional groups such as the Wolof and Maures as well as hybrid groups such as Français-Metiz (French-mixed) and nationalities such as Libyan-Syrians. Hybrid identities and foreign-born were now among the social categories employed.

A concern with where a population originates is an especially impor-tant area of concern for settler societies. The changing nature of and size of subsequent immigrant streams can then be a cause of society. We can get a perspective on the immigrants as a source of anxiety when we consider the national atlases of the United States.

In the United States in 1860 there were five million immigrants and they constituted around 14 percent of the total population. Most immigrants were very similar to the native-born. The economic and political elite were white, Protestant and Anglo and most of the immigrants, but not those shipped over as slaves, were white, northern European and if religious, predomi-nantly Protestant. So, while immigration was an issue, especially the rising number of immigrants from Ireland—the vast majority of whom were not Protestants and many not native English speakers—it wasn't yet seen as an existential threat impacting national identity. The 1874 *Statistical Atlas of the United States* contains data and maps on the foreign-born. There are maps of the foreign-born expressed in both absolute and relative numbers with maps of distinct national groups, including Irish, German, Americans, English and Welsh.

There was mass immigration into the United States from 1880 to 1920. In 1900, the foreign-born had increased to 10 million as a proportion of total U.S. population was at an all-time high. The immigrant streams had changed with more coming from Eastern and Central Europe. Fewer immigrants were English-speaking Protestants. There was growing anxiety at the changing nature of immigrants, especially the growing numbers that were non-Protestant, non-Christian and non-English speaking. There was mounting concern about what this scale of immigration of "others" meant for the character and identity of the nation. Eugenic-inspired texts such as the influential 1916 book *The Passing of the Great Race*, which argued that Nordic people (i.e., white northern European folk) were superior to all others, fueled fears of a weakening of the white race in the United States. The elite concern about the nature of American identity was apparent in the contemporary statistical atlases.

The foreign-born that was a topic of interest in the 1874 atlas became a matter of obsession in the later statistical atlas of 1883, 1898, 1903, 1914 and 1925. These atlases were filled with pages of tables and maps where the population was sliced and diced according to race and immigrant status. Population pyramids were created for the total population of the country and individual states for different groups, including the white population, colored population, native white population, foreign white population, native white of native parents and native white of foreign parents. The married status of single married widowed and divorced, for example, was also produced for total population, native of native parents and native white of foreign parents. There are tables and maps that go into the details of the percentage of foreign-born males aged 21 living in cities with more than 100,000 population. The foreign-born population is used in comparisons with data on disease, crime and public health. Illiteracy was coded by race and inability to speak English. The maps and data all shared the underlying assumption and belief that these new immigrants were a major source of problems. The way that data was presented implied that the new immigrants, because of their genetic inferiority and foreign ways, were undermining the social order.

These atlases were produced and informed by the eugenics movement that was shaped by a toxic combination of elite commentators panicked by the changing nature of the U.S. population and a junk science that claimed to identify distinct ethnic and racial characteristics. A belief in eugenics led to Immigration Acts in 1883, 1885 and 1887 that prohibited Chinese laborers, others from certain countries and "idiots, lunatics and convicts." The

1917 Immigration Act required literacy tests and the 1924 National Origins Quota Act effectively ended immigration, especially those from Asia, Eastern and southern Europe from 1925 until 1965. The act was less about limiting immigration than about maintaining racial purity.[22]

The number of immigrants into the United States fell each year from 1930 to 1970 while immigrants as a percentage of the U.S. population declined steadily from 1910 to 1970. By the time of the 1970 census, the foreign-born constituted less than 5 percent of the total population. The immigration laws became more permissive regarding number and origin after 1965 but the impact was not noticeable until 1980. The 1970 *National Atlas of the United States* was produced at a time when immigration was no longer a hot button topic. The waves of previous immigrants had long been socialized and naturalized. The fear of the foreign other had evaporated. And this change is apparent in the 1970 atlas where the foreign-born is not a category. The socioeconomic section of the atlas portrays a country dominated by its internal movements, rather than external connections. There are maps of population movement within the United States but none that depict the foreign-born or immigrant populations. It is an atlas for a predominant white native-born population slightly isolated from the outside world. That was to change in the next 50 years as immigration increased and the foreign-born became a significant element in the population and the ethnic-racial profile began to change. And with that came a populist and nativist backlash that

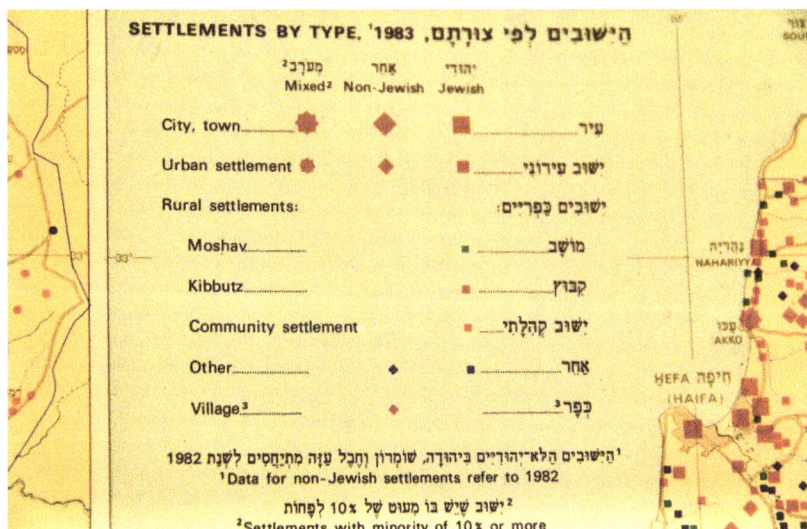

Figure 7.9 Settlement types, *Atlas of Israel*, 1985. *Source:* Photo by John Rennie Short.

became especially visible with the election and presidency of Donald Trump in 2016 on a distinctly anti-immigration agenda.

Israel is a modern-day settler society, and its anxieties of the other also shaped its national atlas. The 1956 *Atlas Yiśra'el* (*Atlas of Israel*) was published in Hebrew. There are maps of Jewish settlements at different dates, of 1931, 1938, 1948 and 1956. After the Six-Day War in 1967 Israel wrested control over the West Bank. The occupied territory became populated with Jewish settlements. Their spread is evident by comparing the 1970 and 1985 editions of *Atlas of Israel* and the 2011 *New Atlas of Israel*. What is of note, in this context, is that the atlases identify distinctly "Jewish" settlements. One map in the 1985 *Atlas of Israel* further categorizes three types of settlement: mixed (with a minority of 10 percent or more), non-Jewish and Jewish (see Figure 7.9). Settlements are recorded around the axis of Jewish/non-Jewish. It is clear the "other" represented in the national atlas of this country is non-Jewish.

To conclude, the national atlas depicts a complex social landscape along ever changing, and sometimes unstable and shifting categories of race, ethnicity, citizenship and identity. The atlases did not simply map the social differences, they created them and, in the process, highlighted the ambiguities, anxieties, fears, paradoxes, difficulties and sometimes the implications and silences of social categorization.

Chapter 8

THE END OF THE NATIONAL ATLAS?

While the title of this book implies the fall of the modern national atlas, this is not a certainty. This chapter's title is more appropriately in the form of a question rather than a definitive statement. It is appropriate then that we consider some of the factors that have almost killed the printed atlas, before looking at new and emerging sources of cartographic vitality. The basic argument of this chapter is that while the old-style national atlas may be dying, new forms may be emerging.

By the end of the short twentieth century, the printed version of the national atlas was collapsing under the weight of its ambition. The 1970 *National Atlas of the United States*—with its 765 maps—weighed in at around twelve pounds. This assemblage comprised prolific efforts of data collection, information assembly, statistical analysis and cartographic display. Eighty-four federal agencies led by the Department of the Interior organized this vast undertaking. Yet only 15,000 copies were printed and distributed, mainly to schools and colleges. The data was quickly outdated and rarely consulted, filling unread, weighty tomes gathering dust on library shelves. It was the last printed national atlas of the United States, soon becoming an historical artifact, reminding us of a different time.[1]

The cosmographers' dream of encompassing the world that lay at the heart of the national atlas was overwhelmed by the accelerating increase of information and data. Heroic attempts to encompass the vast and growing material of a nation included the *National Atlas of India*. The 1959 edition contained 37 maps in one volume. The 1979 edition grew to two volumes while the 1982 edition ballooned to eight. By the time of the 2003–9 printed edition, the national atlas comprised 10 volumes, each 30 cm by 45 cm (roughly 22 by 18 in.). It was more a coffee table than a coffee table book and, possibly, if all the tomes were stacked ingeniously together, a small, intimate coffee shop. The sheer length of the project meant that by the time the volumes were published, much of their data was out of date.

The printed national atlas probably reached its peak around 1960–80 fueled in part by international comparisons. As more nations produced an atlas, pressure grew on those yet to produce one. The 1965 *Atlas Nacional de España* (printed only in Spanish), for example, noted in the introduction that national atlases had been published by Finland, France, Canada, Egypt, Czechoslovakia, USSR, Italy, Australia, Tanzania, Belgium and Israel; national atlases were soon to be completed by Sweden, Poland, India, Morocco, United States, Denmark and Switzerland and they were in various stages of preparation in Bulgaria, Hungary, Indonesia, China, Netherlands, Romania, Turkey and Yugoslavia. Reading the introduction, you get the sense of Spanish desperation and eagerness to join this growing international community. But as the competition to publish national atlases heated up, problems also piled up. The increased bloat of data and the swelling tide of new information made the traditional printed form too expensive, too big and too slow in a fast changing, data-enriched world. The printed national atlas was analog in an increasingly digital world. There were two specific obsolete-making trends: information overload and technological obsolescence. Let's look at each in turn.

Information Overload

Information overload overwhelmed the state's ability to produce data in a timely, cost-efficient and lucid fashion. The time taken to process the information—often going through long periods of production, review and revision—meant that the data rendered meticulously in maps were soon outdated. Expensively produced maps and tables in even the most recently published atlases were soon relegated to historical interest only.

Yet, the national atlas did not disappear. Traditional forms continued to be published in the twenty-first century such as the 2009 *National Alas of Korea* and the *2006 Qatar National Atlas*. Both have sophisticated maps with an interesting and wide range of topics. In the case of Korea, for example, the atlas was published in English and included discussion and maps of environmental/ecological issues and natural disasters. The chapter *Social and Political Geography* comprises maps of labor, gender, public health and social welfare. The *Qatar National Atlas*, published in both English and Arabic, was produced to provide a "general picture of the country to common users within the country and tourists and businessmen visiting the country."[2] It starts off more like the sixteenth-century national atlas of France with a picture of the ruler on the front page, the emir of the State of Qatar, followed by a photographic portrait of the heir apparent. After this obligatory nod to the ruler and to the line of succession, the atlas then takes on a more distinctly modern format with sophisticated data analyses and extensive use of satellite images, including incorporating maps of the labor camps of the immigrant workers who form a transient underclass in the emirate. Figure 8.1 shows a satellite image while Figure 8.2 shows the labor camps; notice some of the larger concentrations numbered 52, 57 and 61.

But these two atlases were the exception rather than the rule. The age of the printed singular national atlas was coming to end as the twentieth century was fading and turning into the twenty-first. One response to the problem of information overload was the splintering of the national atlas into a series of thematic atlases. There were the historical atlases such as the single-volume *Historical Atlas of the United States* (1988) and the three-volume *Historical*

Figure 8.1 Satellite imagery, *Qatar National Atlas*, 2006. *Source:* Photo by John Rennie Short.

Figure 8.2 Labor camps, *Qatar National Atlas*, 2006. *Source:* Photo by John Rennie Short.

Atlas of Canada (1987–93), a massive scholarly work that built on a long and rich tradition of academic historical geography. Both atlases provided exemplary visual accounts of historical change for their respective nations, clearly showing the relationship between geography and history through the clever deployment of maps and figures.

Historical atlases also were produced with a sharper focus on specific events allowing a greater focus and a widening as well as deepening of the range of voices that were heard. An example of a more polyphonic historical atlas is the 2017 *Atlas of the Irish Revolution*, which contains records of the lived experience of a variety of individuals, families and localities involved in the violence of 1916–23. There are more than 300 maps at different scales from nationwide to neighborhood, depicting political support and sites of rebellion. Flow maps show gun-running routes. The maps are contextualized with data, records and copies of documents, official and personal.[3]

In some cases, atlases were produced giving voice to those "nations" previously unheard, silenced or marginalized. The *Stó:lō-Coast Salish Historical Atlas* that came out in 2001was produced with the help of First Nations peoples and gave a cartographic voice to the dispossessed and the formerly marginalized. This atlas was a vehicle for collective identity, the recognition of survival and a claim to legitimacy: an atlas of those facing east and losing much of their land compared to most national Canadian atlases that previously had celebrated those moving west and appropriating the land. Most of the maps in the atlas use indigenous names rather than the Anglo or French names, the dominant toponymic descriptors of the land in the national atlases of Canada.

The subject matter of the traditional national atlas also began to braid into a wide deltaic array of thematic national atlases. The national atlases of China produced at the end of the twentieth century indicate the splintering of atlases to comprehend the vast range of data and issues. A national atlas for China was conceived in 1956 but delayed by the Cultural Revolution: it was printed over thirty years later, but not in the form of a single volume. A population atlas was produced in 1987, a national economy atlas in 1994, a physical atlas in 1999 and two years later, *The Atlas of Population, Environment and Sustainable Development of China*, which included a critical attitude to rapid industrialization and raised environmental issues such as pollution, deforestation, soil erosion and declining air quality.

It was proving impossible to cover the range of available information in a single volume.

The growing data generation and need for analysis prompted more thematic atlases. With social welfare moving up the hierarchy of public demands and government initiatives, countries needed more health data. Figure 8.3 is an example from Mexico's 2002 *Atlas de Salud* (*Atlas of Health*). It was first published in 1993. Atlases of public health were also published in Colombia in 2008, Dominican Republic in 1999, Panama in 1975 and Venezuela in 2003.

The form of these thematic public health atlases also changed. One example is the *Atlas of Variations in Medical Practice in Spain*. Despite the title, this is not an atlas in the traditional sense, but more of a database launched in the

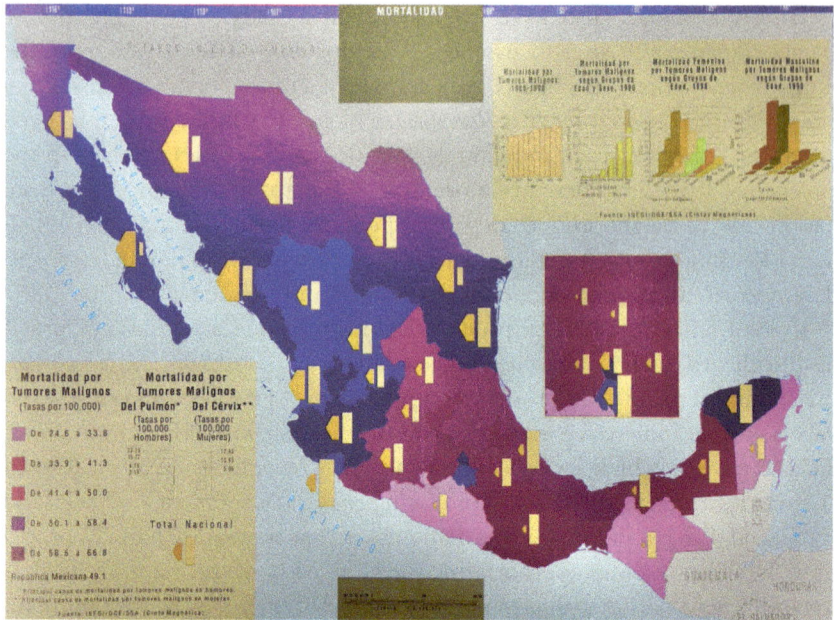

Figure 8.3 Mortality map, *Atlas de Salud* (*Atlas of Health*), 2002. *Source:* Photo by John Rennie Short.

early 2000s by the Spanish National Health System. A group of researchers used this data set to look at over half a million, individual case studies: they mapped the results for different types of care, including spinal fusion and breast surgery. Robust statistical analysis showed that there was systematic and unwarranted sociospatial variation in health care and health outcomes.[4]

Taiwan's 2019 *Diabetes Atlas* is not an atlas in the old, printed form of a book: it is a series of eight papers that mapped diabetes data in Taiwan from 2000 to 2015. The results were published as papers in the *Journal of the Formosan Medical Association*, collectively entitled *Diabetes Atlas, 2019*.[5]

The traditional national atlas was too broad to provide the rich historical details, available in *Historical Atlas of Canada* or the level of granularity of the *Atlas of the Irish Revolution*, too national to allow for the more local nontraditional voices, such as the Stó:lō-Coast Salish people of Canada, and not robust enough to allow for the up-to-date examination of public health issues that was available in the Spanish or Taiwanese medical atlases.

New Technologies

The second trend that made the traditional printed atlas obsolete was the radical and relatively sudden change in how information was processed, displayed and transmitted. Initially, the shift was away from the printed atlas to the digital atlas.

In some cases, this simply involved a change from printed pages to digital files. The 2004 *Atlas of Switzerland*, available in English, French, German and Italian, was produced as a read-only CD-ROM. The 1997–2014 *National Atlas of the United States* was available online only until it too was removed from service in 2014 and merged with the National Map, a website providing mainly topographic data.[6]

The internet rendered the printed national atlas too expensive and too old-fashioned a system to transmit information. The internet allowed greater access to data sources. Cheaper and more readily available computers and sophisticated programs also widened access to data analyses and map production. Atlases and mapmaking were moving to the web.[7] Consider the case of the Dutch national atlas. The first was published in 1963–78. A second edition was published in 1985–95 by the Atlas Bureau of the Ministry of Education. After the completion of the second edition, the bureau was closed. The Foundation for the Scientific Atlas of the Netherlands placed all the maps from the two editions on the web.

Widely available geographic information systems (GISs) allowed more people to assemble, interrogate and display data in real time at low cost. Unlike the stasis of the maps in the printed national atlas, GISs allowed easy and constant updates. Printed maps can show many things. Flow maps, for example, give an indication of connections and movement, yet they freeze the movement. The flows remain as fixed lines on a map. A sense of movement and change is made possible through presenting snapshots in a historical sequence of maps, but even more so in a dynamic cartography that allows flows to actually flow across the animated screen.

Cartographic practices and spatial awareness changed over the twentieth century. William Rankin describes a shift, in part promoted by economic globalizations and military technical developments, from terrain-based maps to the grid of latitude and longitude to a world encased by GPSs.[8] The move from maps of bounded territories to points in a global network reflected our understanding of the world and our ideas of globalization. Manuel Castells wrote about visualizing the world less as a space of places and more as a space of flows.[9] This new spatial awareness embodies new ways of understanding of the world that undermines the insular assurance and territorial fetishization of the national atlas project. New cartographic practices embody new epistemologies, new ways of understanding the world in which the nation-state is neither so central nor so singularly important.

National Atlas as Digital Portal

While the traditional printed national atlas may be dead or dying, the national atlas may be reimagined as a digital portal. Although more information and the internet signaled the end of the traditional national atlas, it also raised the

Table 8.1 Digital Geo-information.

Earth Observation Data
https://www.usgs.gov/centers/eros/data-toolsEarthworks
https://earthworks.stanford.eduFAO GeoNetwork
http://www.fao.org/geonetwork/srv/en/main.homeFrance geoportal
https://www.geoportail.gouv.frGIS Data Sets
https://www.diva-gis.org/DataGlobal Administrative Area
https://gadm.orgGlobal Land Cover
https://www.usgs.gov/media/images/global-land-cover-characteristics-data-base
    -version-20International Climate Data
https://iridl.ldeo.columbia.eduNASA Data Portal
https://nasa.github.io/data-nasa-gov-frontpage/Natural earth
https://www.naturalearthdata.comOpenStreetMap data
https://www.openstreetmap.org/#map=4/38.01/-95.84https://download.geofabrik
    .deWeather Datahttps://www.ncdc.noaa.gov/data-access/land-based-station-data
    /land-based-datasets/global-historical-climatology-network-monthly-version-3

possibility of new forms for the national atlas. The national atlas can now be best conceived as a web portal connected to global and national networks of geo-spatial information. A third edition of the Dutch national atlas, for example, is reimagined as a more interactive portal to a range of local national and global data.[10] There is now a range of publicly available data. Table 8.1 lists just some of the global data sets that can be accessed easily. The national atlas can thus be imagined as part of a global, geo-spatial data infrastructure. The traditional printed atlas had an encyclopedia ambition embodied in a closed text of maps and table. A digital atlas, as an open and interactive text, allows the easier juxtapositions of satellite images, photographs, tables and figures and a wide range of constantly updated data. The shift to a more interactive portal also allows a shift in map production from supply driven to demand driven and from professional cartographers to a range of users.

Toward a Polyphonic Participatory Atlas

As someone with a long fascination for the national atlas, I must admit that the demise of its printed version and even the decline of the digital in some countries such as the United States make me somewhat sad. To be sure it was always a flawed vehicle. In its traditional form, it gave voice to only certain topics and a platform for only certain types of people; its partial nature often glossed over by claims to objectivity. Yet it is too easy to dismiss the national atlas as simply part of an ideological state apparatus to ensure legitimacy and maintain control. At its very best, the national atlas was a cartographic covenant between the state and the people signaling science, embracing progress, suggesting national coherence and embodying national pride. It also displayed the capacity to be more critical. National atlases of China and Vietnam, for example, produced under authoritarian regimes, provided platforms for critical voices questioning the reckless pursuit of economic growth at all costs. The postcolonial national atlas provided some of the most trenchant critiques of colonialism, often with an early critical reckoning. It would go too far to say that the national atlas was a radical document, but it would also be mistaken to reduce it simply to government propaganda. The national atlas had enough "space" to enable alternative, even very critical, voices to be heard. It always had the promise of being a more open than closed text.[11] The national atlas was always limited as a purely ideological instrument of the state. The wishes of the state are often undermined at times by the discourses of science and social awareness. It was never all that effective as part of the ideological state apparatuses anyway. It was too intellectual, too influenced by other discourses and too limited in its consumption. The national atlas was a state document overlain with discourses of science and social investigation, which meant it had the capacity to provide sober assessment not simply national celebration and to be critical as well as laudatory, questioning rather than simply affirming.

The new technologies of data assembly and map production portend a decline of the singular national narratives of the traditional printed national atlas. But they also provide an opportunity for a new form of national atlas. The printed atlas was a fixed text, assembled and narrated by "experts." A new form of the national atlas may be reimagined as an interactive portal with space for different voices to be heard and for new ideas to be quickly introduced. We can reimagine the national atlas along the lines of a giant Wikipedia entry, moderated but crowdsourced and open to constant debate and improvement, ever exposed to global concerns, national issues and more local perspectives. The traditional national atlas displayed a kind of "spatial fetish," a reification of the national territory. But any region or territory is also part of wider linkages and flows. An important role for any postmodern, polyphonic interactive, participatory national atlas would be to show how the region under consideration is both connected and disconnected to wider regional and broader global links and flows. Reimagined as a participatory portal, and as a vital element in a more critical cartography, a national atlas could more easily allow the expression of many voices, different experiences and competing narratives.[12]

The chapter's title posed the question of if this is the end of the national atlas. There are two answers. On the one hand, the printed national atlas is coming to the end of its "natural" life—a life lived in the time of print and in an age of more national optimism and vigorous national science. It was a vehicle capable of change as it encompassed a wide arc of subjects and could change focus and emphasis. The environment, for example, was increasingly portrayed less as an empty container of resources to be exploited and more as a living organism to be managed and protected to ensure a sustainable future. The population was reimagined from subjects to citizens. These innovations finally splintered into more thematic atlases. On the other hand, new technologies could be the beginning of a new era for a reimagined national atlas as a web-based portal to a wide range of geo-information continually updated, interactive and participatory. This new national atlas will be less of a fixed, closed text and more of an open text capable of creative interaction: a portal, with huge educational opportunities to create a socially critical, empirically grounded and theoretically informed understandings of a nation and the world.

NOTES

1. Introduction

1 N. Monmonier, "The Rise of the National Atlas," in *Images of The World: The Atlas Through History*, ed. J. A. Wolter and R. E. Grim (Washington, DC: Library of Congress, 1997), 369–400. Quote from p. 369.

2 One reviewer suggested that a more precise defiinion could be what Eric Hobsbawm descibed as the "short twentieth century" from 1914 to 1991 that followed the "long nineteenth century" from 1789 to 1914.

    E. Hobsbawm, *The Age of Extremes: The Short Twentieth Century, 1914–1991* (London: Michael Joseph, 1995).

3 J. Branch, *The Cartographic State: Maps, Territory, and the Origins of Sovereignty* (Cambridge: Cambridge University Press, 2014).

4 Timothy Mitchell uses the term with reference to British colonization of Egypt. The idea can be extended to other forms of colonization, both political and ideological, as well as to distinctly national projects.

    See T. Mitchell, *Colonising Egypt* (Cambridge: Cambridge University Press, 1988).

5 J. B. Harley, *The New Nature of Maps: Essays in The History of Cartography* (Baltimore and London: Johns Hopkins University Press, 2001).

6 B. Anderson, *Imagined Communities: Reflections on the Origin and Spread of Nationalism* (London: Verso, 2006, revised edition).

7 The connection between national identities and the discourse of geography is considered by a number of authors; see, for example, D. Hooson, ed., *Geography and National Identity* (Oxford: Blackwell, 1994); A. Paasi, *Territories, Boundaries and Consciousness* (Wiley: Chichester, 1996).

8 Thongchai Winichakul develops the idea of *geo-body* in his interesting study of Siam where new cartographic methods were adopted in the late nineteenth century. Although the country had its own long established methods of mapping, they were not legible to European powers. Siam had to adopt Western cartography to negotiate its territorial claims with Western powers. What he terms the *geo-body* of the nation emerged in the new maps made after this transition.

    T. Winichakul, *Siam Mapped: A History of the Geo-Body of a Nation* (Honolulu: University of Hawai'i Press, 1994).

9 J. R. Short, *Imagined Country: Society, Culture and Environment* (London and New York: Routledge, 1991).

10 *Oxford English Dictionary* 1961, Volume XI, p. 215.

2. The Early National Atlas

1   I have developed this theme in a previous book and draw heavily from it here.
     J. R. Short, *Making Space: Revisioning the World, 1475–1600* (Syracuse: Syracuse University Press, 2004).
2   B. Schmidt, *Inventing Exoticism: Geography, Globalism, and Europe's Early Modern World* (Philadelphia: University of Pennsylvania Press, 2015).
3   I. C. Cunningham, ed., *The Nation Survey'd: Essays on Late Sixteenth-Century Scotland as Depicted by Timothy Pont* (Edinburgh: Tuckwell Press, 2001).

3. Cartographic Anxieties and the Emergence of the Modern National Atlas

1   S. Krishna, "Cartographic Anxiety: Mapping the Body Politic in India," *Alternatives* 19 (1994): 507–521. The phrase was also invoked in D. Gregory, *Geographical Imaginations* (Oxford: Blackwell, 1994) and further developed in J. Painter, "Cartographic Anxiety and the Search for Regionality," *Environment and Planning A* 40 (2008): 342–361.
2   M. Arana, *Bolivar: American Liberator* (New York: Simon and Schuster, 2013).
3   An excellent introduction to the cartographic history of Latin America is J. Dym and K. Offen, eds., *Mapping Latin America: A Cartographic Reader* (Chicago: University of Chicago Press, 2011).
4   E. Sanchez, "Augustin Codazzi 1793–1859," in *Geographers Bibliographical Studies*, ed. T. W. Freeman, vol. 12 (Mansell: London, 1988), 35–46. See also L. D. Castillo, "Interior Designs," in *Mapping Latin America: A Cartographic Reader*, ed. J. Dym and K. Offen (University of Chicago Press, 2011), 148–152.
5   M. M. Carrera, *Traveling from New Spain to Mexico: Mapping Practices of Nineteenth-Century Mexico* (Durham and London: Duke University Press, 2011).
6   R. B. Craib, "Historical Geographies," in *Mapping Latin America: A Cartographic Reader*, ed. J. Dym and K. Offen (University of Chicago Press, 2011), 153–158.
7   Carrera, op. cit. Quote from p. 225.
8   Survey of Israel, *Atlas of Israel* (Jerusalem Ministry of Labour, 1970). Quote from p. vii.

4. Cartographic Ruptures and the National Atlas

1   S. Jaatinen, "The National Atlases of Finland," *GeoJournal* 6 (1982): 201–208.
     K. Kosonen, "Maps, Newspapers and Nationalism: The Finnish Historic Experience," *GeoJournal* 48 (1999): 91–100.
     K. Kosonen, "Making Maps and Mental Images: Finnish Press Cartography in Nation-Building, 1899–1942," *National Identities* 10 (2008): 21–47.
2   Address of the President of the United States, "Delivered as a Joint Session of the Two Houses of Congress," January 8, 1918. https://babel.hathitrust.org/cgi/pt?id=mdp .39015074797914&view=1up&seq=1
3   S. Seegal, *Map Men: Transnational Lives and Deaths of Geographers in the Making of Eastern Central Europe* (Chicago: University of Chicago Press, 2018).
     S. Seegal, *Mapping Europe's Borderlands: Russian Cartography in the Age of Empire* (Chicago: University of Chicago Press, 2012).

4   F. Ratzel, "Lebensraum: A Biogeographical Study [2018]: [translated into English by Tul'si (Tuesday) Bhambry]," *Journal of Historical Geography* 61 (1901): 59–80.

5   E. Romer, *Geograficzno-statystyczny atlas Polski* (Warsaw and Krakow: Gebethner and Wolff, 1916).

6   M. Labbe, "Eugene Romer's Atlas of Poland: Creating a New Nation State," *Imago Mundi* 70 (2018): 94–113.

7   R. J. Moore, "Poland's Declaration of Independence: Eugeniusz Romer and His 1916 *Atlas of Poland*," *The Portolan* 6 (2016): 39–52.

8   V. I. Lenin, *Report on the Work of the Council of People's Commissars*, 1920. http://soviethistory.msu.edu/1921-2/electrification-campaign/communism-is-soviet-power-electrification-of-the-whole-country/

9   M. Bassin, *Imperial Visions: Nationalist Imagination and Geographical Expansion in the Russian Far East, 1840–1865* (Cambridge: Cambridge University Press, 1999).

10 Editura Academiei Republicii Socialiste România, *Atlas Republica Socialistă România* (Bucureşti: Academiei Republicii Socialiste România, 1974), vi.

11 Organization of Surveying and Cartography, *National Atlas of the Democratic Republic of Afghanistan* (Warsaw: Geokart, 1985), vii–vii.

12 Op. cit., viii.

13 Ethiopian Mapping Authority, *National Atlas of Ethiopia* (Addis Ababa: People's Republic of Ethiopia, 1988), iii.

14 Instituto Cuban de Geodesia y Cartografia, *Atlas nacional de Cuba* (La Habana: Academia de Ciencas de Cuba, 1970). Cited quote from p. iv is my translation.

15 G. F. Cram, *Imperial Atlas of the Dominion of Canada and the World* (Toronto: Arnt-Gill Company, 1905), 179.

16 Geographical Branch, *Atlas of Canada* (Toronto: Macmillan, 1974), iii.

17 Surveyor General of Pakistan, *Atlas of Pakistan* (Rawalpindi: Survey of Pakistan, 1997), i.

18 Survey of Kenya, *National Atlas of Kenya* (Nairobi: Survey of Kenya, 1970), 82.

19 National Environment Agency, *Vietnam National Atlas* (Hanoi: Ministry of Science, Technology and Environment, 1996), v.

20 M. Najgrakowski, ed., *Atlas Rzecypospolitej Polskei* (Warsaw: Glowny Geodeta Kraju, 1993). Quote from unpaginated sheet.

5. National Atlas, Global Discourses

1   J. R. Short, A. Boniche, Y. Kim, and P. Li, "Cultural Globalization, Global English and Geography Journals," *Professional Geographer* 53 (2002): 1–11.

2   Latitude: 56°8'0"N Longitude: 3°50'15"W.

3   E. Danson, *Weighing the World: The Quest to Measure the Earth* (New York: Oxford University Press, 2006).

6. The Physical World of the National Atlas

1   In this section I draw very heavily on my previous work.
    J. R. Short, *Making Space. Envisioning the World, 1475–1600* (Syracuse: Syracuse University Press, 2004).

2   For a range of interpretations, see P. French, *John Dee: The World of an Elizabethan Magus* (London: Routledge, 1972). R. Deacon, *John Dee: Scientist, Geographer,*

Astrologer and Secret Agent to Elizabeth I (London: Frederick Muller, 1968). N. Clulee, *John Dee's Natural Philosophy* (London: Routledge, 1988). E. G. R. Taylor, *Tudor Geography, 1485–1583* (London: Methuen, 1930). A more recent work that puts Dee's work in a wider context is W. H. Sherman, *John Dee: The Politics of Reading and Writing in the English Renaissance* (Amherst: University of Massachusetts Press, 1995).

3   His library included works by Alhazen, Apian, Roger Bacon, Bellaforest, Borrhaus, Camden, Cartier, Columbus, Contarini, Gilbert, Frobisher, Frisius, Mercator, Munster, Fine, Ptolemy, Ramusio, Schoner, Stoeffler, Thevet and Vespucci.

4   Among the many works on Humboldt, see
    D. Botting, *Humboldt and the Cosmos* (New York: Harper and Row, 1973).
    A. von Humboldt, *Views of Nature* (Chicago: University of Chicago Press, 2014).
    A. Sachs, *The Humboldt Current: Nineteenth Century Exploration and the Roots of American Environmentalism* (New York: Penguin, 2007).
    A. Wulf, *The Invention of Nature: Alexander von Humboldt's New World* (New York: A. A. Knopf, 2015).

5   M. Friendly and G. Palsky, *Visualizing Nature and Society*, 2007. https://tinyurl.com/yyg7ob36

6   I draw upon a large body of work. Relevant texts include:
    P. Carroll, *Science, Culture, and Modern State Formation* (Berkeley: University of California Press, 2006).
    R. Drayton, *Nature's Government: Science, Imperial Britain, and the "Improvement" of the World* (New Haven: Yale University Press, 2000).
    S. Jasanoff, ed., *States of Knowledge: The Co-production of Science and the Social Order* (New York and London: Routledge, 2004).
    C. Parenti, "The Environment Making State: Territory, Nature, and Value," *Antipode* 47 (2014): 829–848.

7   *National Atlas of Canada* (Toronto: Macmillan, 1974), vii–viii.

8   *Atlas of Iran White Revolution* (Tehran: Sahab Geographic and Drafting Institute, 1973), 197.

9   *Botswana National Atlas* (Botswana: Department of Surveys and Mapping, 2001), iv.

10 D. Immerwahr, *How to Hide an Empire: A History of the Greater United States* (New York: Picador, 2019).

11 I draw heavily on
    J. R. Short and L. Dubots, "Contesting Place Names: The East Sea/Sea of Japan," *Geography Review*, 2020. https://www.tandfonline.com/doi/full/10.1080/00167428.2020.1827936

12 W. Steffen, J. Grinevald, P. Crutzen, and J. McNeill, "The Anthropocene: Conceptual and Historical Perspectives," *Philosophical Transactions of the Royal Society A: Mathematical, Physical and Engineering Sciences* 369 (2011): 842–867. https://royalsocietypublishing.org/doi/pdf/10.1098/rsta.2010.0327
    W. Steffen et al., "Trajectories of the Earth System in the Anthropocene," *Proceedings of the National Academy of Sciences* 115 (2018): 8252–8259. https://www.pnas.org/content/115/33/8252?mod=article_inline

13 *The Atlas of Population, Environment and Sustainable Development of China* (Beijing: Science Press, 2000), iii.

14 M. I. Gerasimova and M. D. Bogdanova, "Soil Maps in National and Specialized Atlases (Analytical Review)," *Annals of Agrarian Science* 14 (2016): 76–81. http://creativecommons.org/licenses/ by-nc-nd/4.0/

7. The Social World of the National Atlas

1 M. Foucault, *The History of Sexuality Volume 1: The Will to Knowledge* (New York: Random House, 1976 [trans. R. Hurley, 1979]), 137.

2 P. Carroll, *Science, Culture, and Modern Sate Formation* (Berkeley and Los Angeles: University of California Press, 2006).

3 A range of books deal with the impact of the eugenics movement in the United States. See A. Cohen, *Imbeciles: The Supreme Court, American Eugenics and the Sterilization of Carrie Buck* (New York: Penguin, 2016). T. C. Leonard, *Illiberal Reformers: Race, Eugenics and American Economics in the Progressive Era* (Princeton: Princeton University Press, 2017). D. Okrent, *The Guarded Gate: Bigotry, Eugenics and the Law That Kept Two Generations of Jews, Italians and Other European Immigrants Out of America* (New York: Scribner's, 2019).
   The Holmes quote is from *Buck v. Bell* 274 U.S. 200 (1927). https://supreme.justia.com/cases/federal/us/274/200/

4 J. R. Short, J. Vélez-Hagan, and L. Dubots, "What Do Global Metrics Tell Us about the World?," *Social Sciences* 8 (2019): 136. doi: 10.3390/socsci8050136. T. J. Vicino, B. Hanlon, and J. R. Short, "A Typology of Urban Immigrant Neighborhoods," *Urban Geography* 32 (2011): 383–405.

5 T. McCormick, *William Petty: And the Ambitions of Political Arithmetic* (Oxford: Oxford University Press, 2009).

6 B. D, Berkowitz, *Playfair: The True Story of the British Secret Agent Who Changed How We See the World* (Washington, DC: George Mason University Press, 2018).

7 W. Playfair, *The Commercial and Political Atlas and Statistical Breviary* (Cambridge: Cambridge University Press, 2005).

8 The Economist, "Worth a Thousand Words," *The Economist* 385 (2007): 74–76.

9 L. Vaughan, *Mapping Society: The Spatial Dimensions of Social Cartography* (London: UCL Press, 2018).

10 S. Johnson, *The Ghost Map: The Story of London's Most Terrifying Epidemic – and How It Changed Science, Cities, and the Modern World* (London: Penguin, 2006).

11 A. H. Robinson, "The Thematic Maps of Charles Joseph Minard," *Imago Mundi* 21 (1967): 95–108.

12 M.-J. Kraak, *Mapping Time* (Redlands, CA: Esri Press, 2014).

13 E. R. Tufte, *The Visual Display of Quantitative Information*, 2nd ed. (Cheshire, CT: Graphics Press, 2001), 40.

14 G. Palsky, "Connections and Exchanges in European Thematic Cartography: The Case of 19th Century Choropleth Maps," *Belgeo* 3–4 (2008): 413–426. https://journals.openedition.org/belgeo/11893

15 T. Schulz, *The Statistical Atlas: Studies on Classificatory, Conceptual, Formal, Technical and Communication Aspects* (Dresden: Technical University of Dresden, 2014).

16 The atlas is available at the LOC website. https://www.loc.gov/resource/g3701gm.gct00008/?sp=2
   I also draw heavily on my previous work, J. R. Short, *Representing the Republic: Mapping the United States, 1600–1900* (London: Reaktion, 2001).

17 T. Schulz, *The Statistical Atlas: Studies on Classificatory, Conceptual, Formal, Technical and Communication Aspects* (Dresden: Technical University of Dresden, 2014).

18 T. Schulz, "The Statistical Atlases of the Baltic States 1918–1940: The First National Atlases of the Three Newly Independent Countries," *The Cartographic Journal* 55 (2018): 187–195. Quote from p. 194.

19 H. Gannett, *Statistics of the Negroes in the Unites States* (Baltimore: Trustees of the John P. Slater Fund, 1894), Occ. Paper 4. Quotes are from pp. 5, 24–25.

20 *Atlas of the Union of South Africa* (Pretoria: Government Printer, 1960), xxxiii.

21 S, Cohen, *Folk Devils and Moral Panics* (London: MacGibbon & Kee, 1972).

22 D. Okrent, *The Guarded Gate: Bigotry, Eugenics and the Law That Kept Two Generations of Jews, Italians and Other European Immigrants Out of America* (New York: Scribner's, 2019).

8. The End of the National Atlas?

1   T. Widmer, "765 Maps That Drew Us Together," *The Washington Post* B2, December 27, 2020.

2   Qatar Majliis al-Takhtit, *Qatar National Atlas* (Dawhah: Majliis al-Takhtit, 2006), iii.

3   J. Crowley, D. O'Drisceoil, M. Murphy, and J. Borgonov, eds., *Atlas of the Irish Revolution* (Cork: Cork University Press, 2017).

4   E. Bernal-Delgado, S. García-Armesto, S. Peiró, and Atlas VPM Group, "Atlas of Variations in Medical Practice in Spain: The Spanish National Health Service Under Scrutiny," *Health Policy* 114 (2014): 15–30.

5   C. C. Hsu, S. T. Tu, and W. H. H. Sheu, "2019 Diabetes Atlas: Achievements and Challenges in Diabetes Care in Taiwan," *Journal of the Formosan Medical Association* 118 (2019): S130–S134.

6   https://www.usgs.gov/core-science-systems/national-geospatial-program/national-map.

7   B. Köbben and M.-J. Kraak, "Web Mapping," *International Encyclopedia of Human Geography* 2 (2020): 333–337. doi: 10.1016/B978-0-08-102295-5.10565-7

8   W. Rankin, *After the Map: Cartography, Navigation, and the Transformation of Territory in the Twentieth Century* (Chicago: University of Chicago Press, 2016).

9   M. Castells, *The Rise of the Network Society* (Oxford; Blackwell, 1996).

10 B. J. Köbben, "Towards a National Atlas of the Netherlands as Part of the National Spatial Data Infrastructure," *The Cartographic Journal* 50, no. 3 (2013): 225–231. doi: 10.1179/1743277413Y.0000000056

11 M. Dixon and K. Murphy, "Discursive, Persuasive, Informative and Analytical Texts: What's the Real Difference?," *Metaphor* 4 (2019): 41–48.

12 For an introduction to the ideas of critical cartography and counter-cartographies, see J. W. Crampton and J. Krygier, *An Introduction to Critical Cartography*, 2018. https://tinyurl.com/4hdej8v4

K. Orangotango, ed., *This Is Not an Atlas: A Global Collection of Counter-Cartographies*. (Verlag, 2018). https://elibrary.utb.de/doi/epdf/10.5555/9783839445198

APPENDIX: NATIONAL ATLAS BIBLIOGRAPHY

Afghanistan
1985.   *National Atlas of the Democratic Republic of Afghanistan*. Warsaw: Geokart.

Argentina
1886.   *Atlas de la República Argentina*. Buenos Aires: Kraft.
1954.   *Atlas de la República Argentina*. Buenos Aires: Instituto de Geografico Militar.
1962.   *Atlas de la República Argentina*. Buenos Aires: Instituto de Geografico Militar.

Botswana
2001.   *Botswana National Atlas*. Gaborone: Dept. of Surveys and Mapping.

Brazil
1966.   *Atlas Nacional do Brasil*: Rio de Janeiro: Instituto Brasileiro de Geografia e Estatística, Conselho Nacional de Geografia.
2000.   *Atlas Nacional do Brasil:* Rio de Janeiro: Instituto Brasileiro de Geografia e Estatística, Diretoria de Geociências.

Canada
1905.   *The Imperial Atlas of Dominion of Canada and the World*. Toronto: Arnt-Gill.
1906.   *Atlas of Canada*. Toronto: Toronto Lithographing Company.
1916.   *Atlas of Canada*. Ottawa: Department of the Interior.
1958.   *Atlas of Canada*. Ottawa: Geographical Branch.
1974.   *National Atlas of Canada/ Atlas national du Canada*. Toronto: Macmillan.
1987–93. *Historical Atlas of Canada*. Toronto: University of Toronto Press.

China

1987.   *Population Atlas of China.* Hong Kong: Oxford University Press.

1994.   *National Economic Atlas of China.* Hong Kong/New York: Oxford University Press.

1999.   *National Physical Atlas of China.* Beijing: Cartographic Publishing House.

2000.   *Atlas of Population, Environment and Sustainable Development of China.* Beijing: Science Press.

Colombia

1906.   *Atlas completo de geografía Colombiana.* Bogota: Inprenta Electrica.

1967.   *Atlas de Colombia.* Bogota: Instituto Geografico Agustin Codazzi.

1992.   *Atlas de Colombia.* Bogota: Editolaser.

2008.   *Atlas de la Salud.* Bogotá: Imprenta Nacional de Colombia.

Cuba

1884.   *Planos de Comunicaciones de Las Provincias de La Isla de Cuba (Atlas of Communication of the Provinces of Cuba).* Habana: Jose Mendez & Brother.

1898.   *Atlas of the Forts, Cities and Localities of the Island of Cuba.* Washington, DC: War Department Adjutant General's Office.

1949.   *Atlas de Cuba.* Cambridge, MA: Harvard University Press.

1970.   *Atlas de Cuba:* Cambridge, MA: Harvard University Press.

1970.   *Atlas nacional de Cuba.* La Habana: Academia de Ciencias de Cuba: Academia de Ciencias de la URSS.

1978.   *Atlas de Cuba.* La Habana: Instituto Cuban de Geodesia y Cartografía.

1979.   *Atlas demográfico nacional.* Habana: El Instituto.

1989.   *Nuevo Atlas Nacional de Cuba.* La Habana: Instituto de Geografia de la Academia de Ciencias de Cuba y el Instituto Cubano de Geodesia y Cartografía.

1991.   *Atlas socio-económico de Cuba.* La Habana: Instituto Cubano de Geodesia y Cartografía.

Estonia

1925–28. *Eesti Statistiline Album / Estonie Album statistique.* Tallinn: Riigi Statistika Keskbüroo.

Ethiopia

1988.   *National Atlas of Ethiopia.* Addis Ababa: Ethiopian Mapping Authority.

Finland

1899. *Suomen kartasto. Atlas öfver Finland.* Helsingfors: Suomen maantiet-eellinen seura.

1910. *Atlas de Finlande. Soumen Kartasto. Atlas öfver Finland.* Helsinki: Societe de Geographie de Finland.

1925. *Suomen kartasto. Atlas of Finland. Atlas öfver Finland.* Helsinki: Otava.

1960. *Suomen kartasto. Atlas of Finland. Atlas öfver Finland.* Helsinki: Kustannusakeyhtio Otava.

1976–95. *Suomen kartasto. Atlas of Finland.* Helsinki: Maanmittaushallitus: Suomen maantieteellinen seura.

Ghana

1928. *Atlas of the Gold Coast.* Accra: Gold Coast Survey Department.

1945. *Atlas of the Gold Coast.* Accra: Survey Department.

1970. *National Atlas of Ghana.* Accra: Ghana National Atlas Project.

India

1959. *National Atlas of India.* Calcutta: Survey of India Office.

1968. *National Atlas of India.* Calcutta: National Atlas Organisation.

1979. *National Atlas of India.* Calcutta: National Atlas & Thematic Mapping Organisation.

1982. *National Atlas of India.* Calcutta: National Atlas & Thematic Mapping Organisation.

2003–9. *National Atlas of India.* Kolkata: National Atlas & Thematic Mapping. Organisation.

Iran

1973. *Atlas of Iran White Revolution.* Tehran: Sahab Geographic and Drafting Institute.

Israel

1956. *Atlas Yiśra'el.* Yerushalayim: Maḥleḳet ha-medidot, Miśrad ha-ʻavodah.

1970. *Atlas of Israel.* Jerusalem: Survey of Israel, Ministry of Labour.

1985. *Atlas of Israel.* Tel-Aviv: Survey of Israel/New York, NY: Macmillan Publishing.

2011. *New Atlas of Israel.* Jerusalem: Survey of Israel; The Hebrew University of Jerusalem.

Jamaica

1971. *National Atlas of Jamaica.* Kingston: Town Planning Department.

1989. *National Atlas of Jamaica.* Kingston: Town Planning Department.

Japan

1943.   *Atlas of Japan*. Washington, DC: Division of Naval Intelligence.

1977.   *National Atlas of Japan*. Tokyo: Japan Map Center.

1990.   *National Atlas of Japan*. Tokyo: Japan Map Center.

Kenya

1959.   *Atlas of Kenya*. Nairobi: Survey of Kenya.

1962.   *Atlas of Kenya*. Nairobi: Survey of Kenya.

1970.   *National Atlas of Kenya*. Nairobi: Survey of Kenya.

2003.   *National Atlas of Kenya*. Nairobi: Survey of Kenya.

Korea

2009.   *National Atlas of Korea*. Gyeonggi-do, Suwon-si: National Geographic Information Institute, Ministry of Land, Transport and Maritime Affairs.

Latvia

1938.   *Latvijas Statistikas Atlass / Atlas statistique de la Lettonie*. Riga: Valsts Statistika Parvalde.

Liberia

1983.   *Republic of Liberia Planning and Development Atlas*. Monrovia: Ministry of Planning and Economic Affairs.

Lithuania

1918–28. *Lietuva Skaitmenimis—La Lithuanie en chiffres*. Kaunas: Centralinis Statistikos Biuras.

Mexico

1858.   *Atlas geográfico, estadístico, e histórico de la república Mexicana*. Mexico: Imprenta de Jose Mariano Fernandez de Lara.

1885.   *Atlas pintoresco é histórico de los Estados Unidos Mexicanos*. Mexico: Debray Sucesores.

2002.   *Atlas de Salud (Atlas of Health)*. Cuernavaca, Morelos: Escuela de Salud Pública de México: Instituto Nacional de Salud Pública.

Pakistan

1985.   *Atlas of Pakistan*. Rawalpindi: Survey of Pakistan.

1997.   *Atlas of Pakistan*. Rawalpindi: Survey of Pakistan.

2012.   *Atlas of Islamic Republic of Pakistan*. Rawalpindi: Survey of Pakistan.

Peru

1865.   *Atlas Geográfico del Perú*. Paris: Fermin Didot.

Poland

1916. *Geograficzno-statystyczny atlas Polski (Geographical and Statistical Atlas of Poland)*. Warszawa i Kraków: Gebethner i Wolff.

1947. *Studium Planu Krajowego (Studies for the National Plan)*. Warszawa: no publisher identified.

1973–78. *Narodowy atlas Polski (National Atlas of Poland)*. Wrocław: Polskiej Akademii Nauk.

1993. *Atlas Rzeczypospolitej Polskiej (Atlas of the Republic of Poland)*. Warszawa: Główny Geodeta Kraju.

Qatar

2006. *al-Aṭlas al-waṭanī al-Qaṭarī (Qatar National Atlas)*. Dawhah: Majlis al-Takhṭīṭ.

Romania

1974. *Atlas Republica Socialistă România*. Buçureşti: Editura Academiei Republicii Socialiste România.

Russia

2004. *Naṭsional'nyĭ atlas Rossii v chetyrekh tomakh (National Atlas of Russia)*. Moskva: Federal'naia sluzhba geodezii i kartografii Rossii.

Saudi Arabia

1975. *Atlas of Saudi Arabia*. Beirut: Malt International.

1999. *Atlas of the Country of Saudi Arabia*. Riyadh: The Ministry.

Senegal

1925. *Atlas des Cercles de L'A.O.F. Fasicule VII*. Paris: Maison Forest.

1977. *Atlas National du Sénégal*. Paris: l'Institut Géographique National.

Sierra Leone

1953. *Atlas of Sierra Leone*. Freetown: Surveys and Land Department.

South Africa

1960. *Atlas of the Union of South Africa*. Pretoria: Govt. Printer.

Spain

1965. *Atlas Nacional de España*. Madrid: Instituto Geográfico y Catastral.

1992. *Atlas de España*. Madrid: El País-Aguila.

Sri Lanka

1988. *National Atlas of Sri Lanka*. Colombo: Survey Dept.

2005. *Environmental Atlas of Sri Lanka*. Battaramulla: Central Environmental Authority.

2012.   *Population Atlas of Sri Lanka*. Colombo: Department of Census and Statistics.

2016.   *Atlas of Economic Activities of Sri Lanka*. Battaramulla: Department of Census and Statistics.

UAE

1993.   *National Atlas of the United Arab Emirates*. Al Ain, United Arab Emirates/Reading, UK: United Arab Emirates University in association with GEOprojects (U.K.) Ltd.

USA

1874.   *Statistical Atlas of the United States*. New York: Bien.

1883.   *Scribner's Statistical Atlas of the United States*. New York: Scribner's.

1898.   *Statistical Atlas of the United States*. Washington, DC: Govt. Printing Office.

1903.   *Statistical Atlas*. Washington, DC: United States Census Office.

1914.   *Statistical Atlas of the United States*. Washington, DC: Govt. Printing Office.

1925.   *Statistical Atlas of the United States*. Washington, DC: Govt. Printing Office.

1970.   *National Atlas of the United States*. Washington, DC: Geological Survey.

1988.   *Historical Atlas of the United States*. Washington, DC: National Geographic Society.

USSR

1937–39.   *Bol'shoĭ sovetskiĭ atlas mira* (*Great Soviet World Atlas*). Moscow: Scientific Editorial Institute.

Venezuela

1840.   *Atlas Físico y Político de La República de Venezuela*. París: Lith. De Thierry Freres.

1979.   *Atlas de Venezuela*. Caracas: Dirección de Cartografía Nacional.

Vietnam

1909.   *Atlas général de l'Indochine française*. Hanoi: Impr. d' extreme-orient.

1928.   *Atlas de l'Indochine*. Hanoi: Service Géographique de l'Indo-Chine.

1996.   *Vietnam National Atlas*. Hanoi: Ministry of Science, Technology and Environment.

INDEX

www.ingramcontent.com/pod-product-compliance
Lightning Source LLC
Chambersburg PA
CBHW062032270326
41929CB00014B/2412